# OAKS, DRAGONFLIES
# AND PEOPLE

# OAKS, DRAGONFLIES AND PEOPLE

## CREATING A SMALL NATURE RESERVE
## AND RELATING ITS STORY
## TO WIDER CONSERVATION ISSUES

NORMAN W. MOORE

2002

Harley Books (B. H. & A. Harley Ltd.),
Martins, Great Horkesley, Colchester,
Essex CO6 4AH, England

Text set in Sabon

Printed by St Edmundsbury Press Ltd,
Bury St Edmunds, Suffolk

Colour reproduced and printed by
Hilo Colour Printers Ltd,
Colchester, Essex

Bound by Woolnough Bookbinding Ltd,
Irthlingborough, Northants

*Oaks, dragonflies and people*

British Library Cataloguing-in Publication
Data applied for

ISBN 0 946589 71 2

# CONTENTS

# LIST OF ILLUSTRATIONS

COLOUR PLATES (between pages 84 and 85)

1   Aerial photograph of our field, 29 July 1988
2   The Old Enclosure, 1980
3   Lombardy poplars in the Old Enclosure, 1973
4   Meadow Brown (*Maniola jurtina*)
5   Ringlet (*Aphantopus hyperantus*)
6   Gatekeeper (*Pyronia tithonus*)
7   Elms in eastern hedge, July 1973
8   Felling the last but one elm, 1979
9   Large elm felled by great gale of 3 January 1976
10  Dead elms on southern border, 1979, shortly before felling
11  Peter Moore digging holes in Main Plantation, before planting oaks,
       winter 1971/72
12  The Old and Main Plantations, July 1973
13  The Main Plantation, late August 1975
14  The Main Plantation, July 1976 – a drought year
15  The Main Plantation, Autumn 1978
16  The Main Plantation, Winter 1980
17  The Main Plantation, Autumn 1981
18  Path through Main Plantation, Winter 1990/91
19  Cowslips among Cow Parsley and Cleavers, 1998
20  Daffodils in Old Plantation, 1990
21  The Corner Plantation, April 1998
22  Digging the Mere, December 1983
23  Shelter provided by spoil heap, Spring 1984
24  The Mere's eastern shelter belt, 1998
25  The Mere filling up, Spring 1984
26  The Mere becoming established, August 1984
27  The Mere at two years ideal for dragonflies, Summer 1985
28  The Mere following drought, February 1995
29  The Mere showing encroaching reeds and bulrushes, February 1995
30  The Mere Meadow, April 1998
31  The Rookery, Buckingway Plantation, April 1998

TEXT FIGURES

LIST OF TABLES IN TEXT

To my grandchildren

# PREFACE

NATURE CONSERVATION has been a concern of mine since I was a child and it has been at the centre of my working life since the early 1950s. I have described some of my experiences in *The Bird of Time*. In that book I showed a sketch of a pond and its surrounding rough grassland which were parts of a nature reserve that I had created in the field at the back of our house in Swavesey in Cambridgeshire. The wood was mainly for birds, the grassland for butterflies and the pond for dragonflies, my favourite insects. My small reserve has given me great pleasure for nearly forty years. It has influenced what I have tried to do in my professional life as a conservation biologist, and it has provided me with food for thought about the big conservation issues which we all face today.

In the first part of this book I try to share my pleasures in creative conservation by describing how problems were solved in setting up this reserve. I describe how the different parts of it were colonized by a wide range of birds, mammals, butterflies, dragonflies and other animals. In the second part, I discuss general conservation problems which I found were illuminated by my experience in creating woodland, grassland and wetland habitats. I ask to what extent such created habitats *can* be used as substitutes for ancient woods, well-established meadows and water bodies. This is a question of special interest to anyone with a fairly large garden, to farmers and to organizations and businesses of many kinds. This topic leads me to ask what is the role of nature reserves in the new millennium. I return to the theme of *The Bird of Time* – that conservation is about the future and so time must be taken into account when determining conservation strategy for mankind and his environment. In particular, I consider the particular difficulty parliamentary democracies have in dealing with the long term. I conclude by making suggestions on how to overcome this daunting problem.

# ACKNOWLEDGEMENTS

It is a great pleasure to thank all those who have helped me create my reserve at Swavesey and to write this book about it.

First I should like to thank all my ex-colleagues on the staff of the Nature Conservancy Council who gave me the wherewithal to create the Mere in 1983, and especially Dick Steele, who earlier had provided the young oak trees which formed the basis for the Main Plantation. I am grateful to the many others who have given me young trees and other plants for the Wood and the Mere. I should like to thank my farming neighbour John Burgess for all the help he has given me, in particular for fencing the Mere Enclosure and the Corner Plantation. I am grateful to all my neighbours for so kindly tolerating the changes which I have made on the land adjoining their own.

In writing this book I have sought specialist advice from several people. I should like to thank especially Max Walters for advice on plants, Beth Davis for advice on archaeological matters and John Shepperson for advice about the Swavesey Enclosure Act and the Swavesey Tythe Map. I thank the Committee for Aerial Photography of the University of Cambridge for permission to publish aerial photographs of Swavesey, and the Ordnance Survey for permission to base sketch maps on their Ordnance Survey maps. I thank Ken Dolbear for the photographs of dragonflies and Ken Willmott for the photographs of butterflies. I also thank Gerald Fleuss for redrawing my text figures. All other photographs and drawings were mine.

This book has been greatly improved by suggestions made by my publishers Basil and Annette Harley, who are themselves enthusiastic naturalists and conservationists. Above all I wish to thank my wife Janet Moore. Her lifelong experience in correcting biology students' essays has made her suggestions enormously valuable. I am greatly indebted to her for the constructive scrutiny she has given every chapter of this book. It is a pleasure to thank my friend Zoë Conway Morris for typing the several drafts of this book so painstakingly. I am most grateful to her for her patience and skill and for the valuable suggestions which she has made in producing it. James Shurmer has skilfully achieved a clear and suitable design and Susan Williams has helped with the preparation of the index.

Finally I should like to thank Janet again and our children Peter, Caroline and Helena and our eight grandchildren for tolerating the time I have spent on the reserve over the years, and for giving their practical support and encouragement, without which nothing would have been achieved. It is a great pleasure that they all enjoy the results of our labours.

# PART ONE

## INTRODUCTION

IT rained and rained and rained. We were told that it was the wettest summer in Chile for years. That year (1991), Janet and I had come to study dragonflies of the Neopetaliidae, a Gondwana family which is confined to Australia and Chile. Although Australia and Chile have been separated for over forty million years these striking dragonflies with red spots on their wings are remarkably similar in both countries. We were also searching for the larva of a huge black, primitive dragonfly called *Phenes raptor* which up to that time had never been seen by entomologists.

One day we ventured out and got soaked as we watched a young black-crowned night heron cowering by a swollen, turbulent stream. The weather was hopeless for dragonflies and precious little use for birds. We took the hint, and returned to the little house at the foot of the mighty, but totally invisible Choshuenco volcano, where we were staying. We watched our host's family prepare for Christmas, read paperbacks and listened impatiently as the rain beat on the corrugated iron roof. I thought of home nearly eight thousand miles away and about the ways I had changed our field, and about the plants and animals which had come to live in the habitats I had created. Distance from home seemed to make it all feel much more interesting. Might it even be worth writing a book about my experiences?

Since 1961 we had owned an 8-acre field adjoining our garden. It was a very ordinary grass field, very flat, in an area devoted almost entirely to arable farming. It was partly bordered by other people's gardens, a ditch unromantically known as Public Drain Number 12, which was usually dry, and a strip of waste ground that eventually became a site for mobile homes. To say the least it was a challenge for a conservationist, but of course I had to take it up. In the intervening years I have planted the first wood to have existed in the district for many years, and have constructed a pond which is now one of the best ponds for dragonflies in Cambridgeshire. I have had immense pleasure in doing these things and I have learnt a great deal in the process. As the rain pelted down in the foothills of the Andes I became more and more convinced that there was a story worth telling and that it had implications which were worth exploring. I decided to start writing a book when we got home.

My good intentions were not fulfilled quickly. When we got back to England there was much to do – helping to look after Wicken Fen for the National Trust, organizing the conservation work of the British Dragonfly Society, writing the Dragonfly Action Plan for the Species Survival Commission of the IUCN and much else. The book progressed very slowly. It was not until 1997, when I had been relieved of some of my conservation duties by kind colleagues, that I was able to start again with this story about conservation in a field in a Cambridgeshire village.

# CHAPTER ONE

# WE BUY A HOUSE AND THEN A FIELD

IN 1959 the Nature Conservancy made plans to establish an experimental
station to undertake research on conservation problems. It was to be in the
west of East Anglia so that use could be made of the cluster of National
Nature Reserves near Peterborough: Monks Wood, Woodwalton Fen, Holme
Fen and Castor Hanglands. Eventually a site for the station was chosen in a
field on the edge of Monks Wood. I had been asked to head a team to study
the effects of pesticides and other pollutants on the environment and
after some arm-twisting by Max Nicholson, our Director General, and Barton
Worthington, our Scientific Director, I had accepted. At the time I was the
Nature Conservancy's Regional Officer for South-West England and was
based at Furzebrook Research Station near Wareham in Dorset. So we had to
move house from Dorset to East Anglia. I spent much of 1960 writing up my
research work on Dorset heathlands and looking for a house which was acces-
sible both to Monks Wood and to Cambridge where our children hoped to go
to school and where Janet could return to academic life.

There was little new building in west Cambridgeshire at that time, and few
older buildings were for sale. But eventually I came across an advertisement
for 'A Pleasantly Situated Country Property Known as the Farm House' in
Boxworth End in the southern part of Swavesey. It lay 17 miles from Monks
Wood and 8 miles from Cambridge. It looked just what we wanted and I
was very disappointed when I called on the owners to be told that it was about
to be sold to someone else. However there were difficulties because the
prospective buyer had discovered that the wall which formed the northern
boundary of the garden was in danger of collapsing and falling into Public
Drain Number 12. Accordingly he insisted that the asking price of £4,250

should be reduced. Having looked at the house and the wall I was prepared to accept the original price of the house and the hazards of the wall and so the owners agreed to sell The Farm House to me.

We sold our house in Dorset and I arrived at our Cambridgeshire farm house on 28th August 1960 with as much luggage as our ancient Morris Traveller could hold, which included our son's budgerigar in its cage. On 31st August I met my family – complete with guinea-pigs – off a steam train at Swavesey station and we settled in.

The Farm House is typical of many farm houses in the district which were built at the time of the Enclosures between 1820 and 1840. It is made of the yellow-grey brick characteristic of the area and has a slate roof. Two large lime trees, a yew tree and the old fruit trees in the little orchard and garden are probably about the same age as the house. The garden looked a suitable habitat for common garden birds but the nearness of the farm buildings to the south and east meant it was the happy chirping ground of enormous and perhaps excessive numbers of house sparrows.

We had hardly settled in to the Farm House when we learnt that the pasture field of 7³/₄ acres the other side of the collapsing wall would soon be up for sale (Figure 1). It belonged to a builder who had recently bought it from Mr Northrop, our neighbour, who had a small farm the other side of the road. Unfortunately for the builder, but fortunately for us, he could not get permission to build because our end of the village had not yet acquired main drainage, and he could not wait until it was installed.

We were faced with a chance we dare not miss, despite the state of our bank balance. If we did not buy the field we would probably find ourselves eventually bordering a housing development undertaken by another builder or a busy road leading to one. If we did buy the field we could fulfil our elder daughter's often expressed wish to keep a pony and, if I were ever to have the time in the future, I could indulge in some practical nature conservation. We decided to try and buy the field.

The builders had offered to sell the field back to Mr Northrop but he did not want to buy it as he planned to retire shortly. However, until he retired he did want to continue to graze his cattle in the field. So we agreed that if I could get the £1,000 necessary to buy the field I would buy it through Mr Northrop and he would rent the grazing for as long as he needed it. We borrowed the £1,000 from a kind cousin who earned rather more as a business man than I did as a government scientist. Although the conveyancing of the field took a long time to complete, eventually it was ours. It was well worth waiting for.

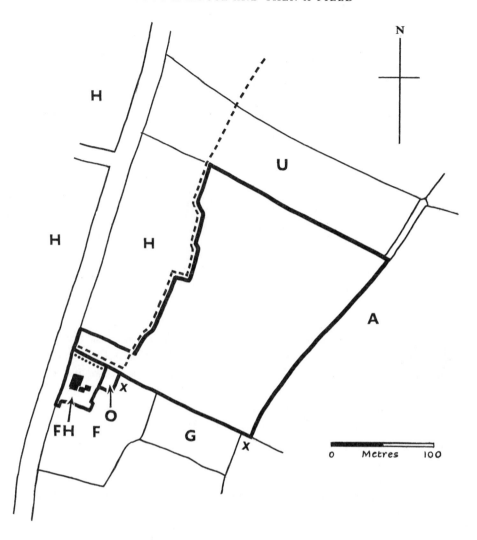

Figure 1: Our field and its surroundings in Boxworth End, Swavesey, 1961

| | |
|---|---|
| ——— | external and internal boundaries of our field, house and garden |
| ——— | other boundaries |
| - - - - - | Public Drain Number 12 |
| ·········· | collapsing wall |
| A | arable field |
| F | farm buildings and yards |
| FH | The Farm House and its garden |
| G | permanent grass |
| H | houses and gardens |
| O | our orchard |
| U | uncultivated grassland (later Pine Grove Caravan Park) |
| X–X | south border of our field, a fence lined by large elm trees |
| | For details in 1998 see Figure 6 and inside front cover |

# CHAPTER TWO

# SURROUNDINGS – SWAVESEY, A FEN EDGE VILLAGE

O UR field lies near the centre of the parish of Swavesey which I shall describe because it provides not only the location of our field but also the sources from which it has been colonized by plants and animals. Many of the locations mentioned below are shown on the map (Figure 2).

Swavesey is one of the dozen or so villages which lie on the rather featureless clay plain south of the Fens and north of the Roman road between Cambridge and Godmanchester. This road is now the A14, a major route linking the east coast ports with the Midlands.

Today Swavesey is a large and growing village of 2,000 inhabitants with several shops, two pubs and some light industry. Bumps and hollows in the ground at both ends of this long village suggested it had an interesting past. It has, as the late Jack Ravensdale showed in his book, *History on your Doorstep*. The irregularities in fields and orchards about 500 yards north and west of our field indicate the site of the original village (Figure 3). Archaeological remains and documentary evidence show that in the Middle Ages a new settlement was later established a mile to the north on the fen edge. The bumps and depressions which can be seen by the church are the remains of a monastery which was set up in the thirteenth century, and which in the following century became a Carthusian convent until the dissolution of the monasteries in 1539. The little hillock in the north-west of the village is the motte of a motte-and-bailey castle. The moat round the bailey still contains water and badgers made a sett in the earthworks. The castle protected the entrance to the new settlement, which was in fact no less than a planned and fortified mediaeval town. It was established by the de la Zouche family.

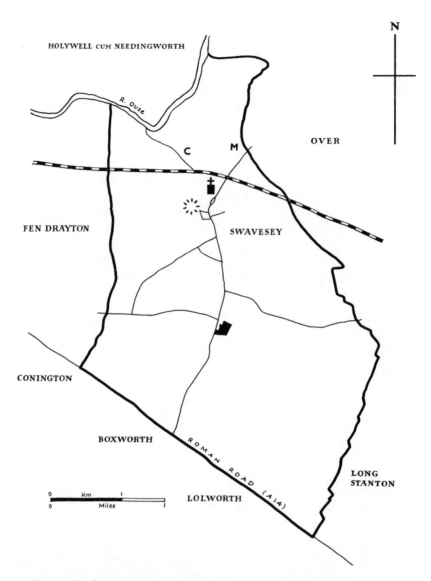

Figure 2:  The parish of Swavesey and the position of our field in it, 1961

Swavesey lies between the Great Ouse River to the north and the Roman Road between
Godmanchester and Cambridge to the south. Our field is shown in black.

The abandoned mediaeval village (see p.16) lay just to the west and to the north of our field.
Since the 13th century the village has become centred on the north of the parish. The church and
site of the Priory are indicated by a ⚑ and the site of the castle by ⁖. C is the Navigation Drain,
the disused canal. M is Mare Fen, which is bordered on the east side by the Swavesey Drain.
Metalled through-roads within the parish are indicated, together with the railway (now disused).
Most of the parish north of the railway consists of grassland liable to flooding. Today the
western part of this area contains two large water-filled gravel pits. Most of the parish lying
south of the railway consists of arable land. Of the surrounding parishes only Boxworth and
Lolworth have names indicating clearances from forest (see p. 20).

© Crown copyright reserved

Figure 3: Aerial photograph taken on 30th June 1987 showing our field and its sur-roundings at Boxworth End, Swavesey

Our field is the light coloured one at the bottom right of the photograph. The planted part of it appears dark. For details, see Plate 1 and Figures 1 and 6. The first Swavesey settlement lay to the west and north of our field. Field marks indicating the northern part of this ancient village can be seen in the dark field at the top right of the photograph just north of the Dairy Farm (top right).

For many years the prosperity of Swavesey was due to its role as an inland port. Its docks survive as ponds. However, when the railway came in 1847, the canal (C) which had linked Swavesey to the River Ouse went out of use. Swavesey lost its original *raison d'être*.

The main land-use in Swavesey has always been agriculture. Most of the parish is on Jurassic clay and boulder clay brought down by the Ice Age. The land is hard to work but fertile. In the past there were many more farm animals in Swavesey than today. There was no piped water so cattle and horses depended on ponds. At the time of the Swavesey enclosures (1838) there were twenty ponds within 500 yards of our house. Today arable farming predominates and nearly all the ponds have disappeared. There is very little unimproved grassland. The little spinneys present at the time of the Enclosures have disappeared, as have most of the orchards. Many hedges were taken out during or after the Second World War; most of those that remain are relatively new and thus are poor in species. Skylarks and corn buntings were typical birds of the area but both have declined catastrophically in recent years. The number of hares fluctuates greatly and rabbits are not abundant. In winter wood pigeons, lapwings and sometimes golden plovers feed in the wide open fields. But in general the southern parts of Swavesey are poor in wildlife.

The northern quarter of the parish is very different. It is flat, low-lying fenland resting on alluvial gravels, sands and peat. Winter floods are frequent and so most of the area is kept as grassland for summer grazing. It is a remnant of a habitat which must have existed over vast areas of the Fens in the past. Several wide droves bordered by hedges run from the fens to the village. They were originally used for driving cattle from the pastures on their way to markets inland. Their upkeep is still maintained by a levy authorized by a special Act of Parliament. Unlike most fens, the Swavesey ones are divided up into fields with tall hedges. They consist mainly of hawthorn, but buckthorn is common in some of them.

The dykes and ditches of the Swavesey fens drain into the River Ouse. Normal flooding is contained by washlands beside the Ouse and the Swavesey Drain which drains into it. Quite frequently in winter the drains and washlands cannot control the floods and the whole area is under water. At such times birds and mammals seek refuge in the hedges including rabbits, which can look uncomfortable and bizarre when perched precariously in hawthorns. In recent years two large gravel pits have been dug in the Swavesey fens together with several more in the neighbouring parish of Fen Drayton. Mare Fen (M) near the site of the monastery has been declared a Local Nature Reserve and its management by the local Wildlife Trust ensures that it is kept wetter than it would otherwise be, to the benefit of wildfowl and skaters.

By contrast with the southern part of the parish, the fenland in its north is rich in wildlife. Butterflies and many other insects breed on the embankments and droves. Many species of dragonflies breed in the River Ouse, and in the gravel pits, drains and ditches. The Ouse is also rich in fish and mollusc species. It is tidal up to Brownshill Staunch just below the Swavesey stretch of the river. This can bring interesting visitors. One winter's day we were walking along the river bank on the parish boundary when we saw two mute swans looking surprised and rather alarmed – not surprisingly because a seal had just surfaced between them. As it was above the lock at Brownshill Staunch it must have got there either by waddling round the lock or by pretending to be a boat.

Snipe, redshank and yellow wagtail occur in the fen meadows. Coots, grebes and a wide range of waterfowl breed in the gravel pits of Swavesey and Fen Drayton, including species typical of the East Coast far to the north such as shelduck, oystercatcher, ringed plover and common tern. In winter the fens and gravel pits are visited by hundreds of wildfowl, notably wigeon. Bewick's swans can sometimes be seen on Mare Fen nearby, together with wigeon, pintail and other species of duck, and at migration time waders appear. They include black-tailed godwits and ruffs, two species which were exterminated in the Fens in the nineteenth century but have returned to breed in the Ouse Washes a few miles to the north east of Swavesey.

The A14, the Roman road which divides the lowland plain on which Swavesey lies from the low hills to the south, makes an important divide. To the north of it there are no woods; to the south of it there are several, and most of the village names there show that the villages were once clearings in the forest: the names end in 'worth' and 'ley' which mean clearings, thus Madingley, Lolworth, Childerley, Boxworth, Elsworth and Papworth. There are no such names in the plain to the north. It seems likely that when the Saxons came the plain beside the fens had already been cleared of forest. Perhaps it is one of the oldest bits of agricultural land in England. This means that when I planted our little wood it was on land that had not seen forest for a very long time, nor were there sources for woodland species nearby other than those which could live in hedges. On the other hand, there were numerous sources for aquatic plants and animals in the northern fenny end of the parish.

# CHAPTER THREE

# THE HISTORY OF OUR FIELD

THE field lies on Ampthill Clay, a Jurassic formation younger than Oxford Clay and older than Kimmeridge Clay. Thus about 150 million years ago our field was part of a great muddy sea. When we dug the Mere in 1983 several shells of a large, heavy extinct oyster (*Ostrea discoidea*) came to the surface. Coming to more recent times, the site of our field must have been covered by ice in the Ice Age. Most of the stones in the gravels in the north of the village are flints, but others appear to be bits of millstone grit from the Pennines and from other rocks from the north. There must have been a period after the Ice Age when our field was under forest, but there is no evidence of what it was like. As mentioned in the previous chapter, the area appears to have been cleared of forest very early on.

When digging the Mere I found signs of human occupation – small bits of crude pottery, shells of the edible mussel (*Mytilus edule*) and animal bones. I sought expert advice on the age of the pottery and was told that it was St Neots Ware from the Romano-British period. The bits included fragments showing the rims of cooking pots. They provide further evidence that the Swavesey area has been inhabited and farmed for a very long time.

When the ditch is cleaned out I often find bricks of unusual dimensions – 8½″ × 4″ × 2″. They are red and must be much older than the yellow-grey brick of our house. They are locally made field bricks – certainly eighteenth-century, very likely a good deal older. The map of the enclosures awarded by the 1838 Enclosures Act clearly shows a quite large L-shaped building on the site where our field joins the road. Later signs of human occupation include bits of Victorian clay pipes and china which I unearth when digging in our garden.

Figure 4: Sketch maps showing changes in our field since the beginning of the nineteenth century

(a) Map based on the first edition of the one-inch (1:63360) Ordnance Survey, 1808–22, revised 1834–35.
What was to become the southern half of our field was the eastern part of an enclosed field or close. It contained a spinney on its northern boundary. To the north of the close the land was part of Swavesey's unenclosed open-field system. It included a line of fruit trees. The small area in the south west corner of our field by the road was largely covered by a building.

(b) Map based on the Enclosure Map of 1838 and the Tythe Map produced a little later.
The southern part of the field was grassland and remained enclosed. It was called Braziers Close. It still contained a spinney. There is still a building in the south west corner of the field by the road. The northern part of what was to become our field has been allotted to Trinity College and has been enclosed. It is called Boxworth End Close. The little building just to the west of Public Drain Number 12 was a dove house.

(c) Map based on the first edition of the 25-inch map (1:2,500) 1878.
The two enclosures shown on the previous map have been amalgamated into one field: what was to become our field was the eastern part of this large field bordering the road and divided by Public Drain Number 12. The building in the south west corner has gone and its site amalgamated with the field. Two ponds are shown in the western part of the field. In the original 1878 map all the elms which were present when we bought the field in 1961 are clearly shown.

(d) The situation when we bought our field in 1961. The map is based on the provisional 2½ inch map (1:25,000), 1945–49 with additions.
Our field is identical to that shown in 1878, but the land between Public Drain Number 12 and the road has been built on and made into gardens except for our bit of it.

(e) Map based on the 25-inch map (1:2,500), 1978.
The field to the north of our field, which was allotted to Mr William Wragg in 1838, has been converted into Pine Grove Caravan Park. The woodland planted by me is indicated on this map.

(f) The situation in 1998. Map based on aerial photographs (Plate 1 and Figure 3).
The Mere, its enclosure and the Corner Plantation have been added to the 1978 map. For more accurate portrayal see Plate 1.

———— external and internal boundaries of our field other than Public Drain Number 12

- - - - - Public Drain Number 12

———— other boundaries on the maps

··········· future boundaries of our field which crossed open land at the time the maps (a) and (c) were made.

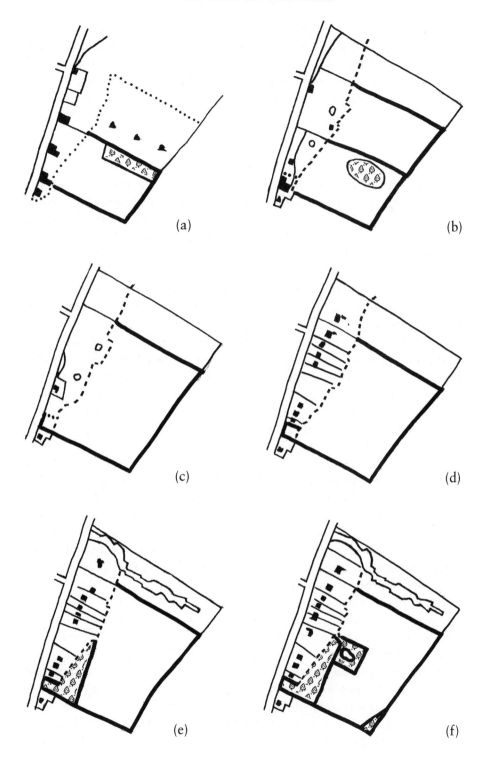

(a)

(b)

(c)

(d)

(e)

(f)

Maps enable us to put dates on events. The first survey for the production of one-inch-to-one-mile Ordnance Survey maps in Cambridgeshire was made between 1808 and 1822. This survey, which contained a number of inaccuracies, was extensively revised in 1834 and 1835 and published in 1836. A revised map with railways added, including the one which went through Swavesey, was published in 1867 and has been reprinted by David and Charles. Unlike later editions of one-inch Ordnance Survey maps it shows field boundaries (Figure 4a). Close examination of old Sheet 51 shows that our field was divided in two at that time: the southern part was a distinct separate field while the northern end was part of an undivided area adjoining the Dairy Farm (see Figure 3). The southern field contained a small rectangular wood. Today this part of our field is traversed by two east-to-west depressions indicated by arrows (Plate 1). This suggests that, before the production of the first Ordnance Survey map, the original field may have been subdivided into three small fields or closes.

The Enclosure map of 1838 and the Tythe map, which was produced a little later, provide a great deal of information (Figure 4b). At this time, about 160 years ago, what was to become our field had been the eastern parts of two fields adjoining the road – that is the parts of the fields east of the Public Drain. In the document accompanying the Tythe map, the southern field was called Braziers Close. It was the field outlined in the first Ordnance Survey map (see Figure 4a). At the time of the Tythe map it was grassland and it still contained a wood. In the same document, this is called a spinney and is shown as roughly oval in shape and not contiguous with the northern boundary of the field. It is not obvious whether the shape and position of the spinney had changed in the short period between the production of the Ordnance map and the Enclosure Award and Tythe maps, or whether the Ordnance map was inaccurate. Just conceivably the field's name derived from the fact that the field contained a brazier for making charcoal from the wood of the spinney. More likely Brazier was someone's name. The interesting thing is that our field did once contain a small wood, although today no signs of it can be seen on the ground nor on aerial photographs. Braziers Close was owned by Trinity College, Cambridge, and farmed by their tenant, Mr Joseph Carter.

The northern part of what was to be our field, which had been part of the ancient open-field system, had been allotted to 'The Master, Fellows and Scholars of Trinity College' under the Enclosure Act and, like Braziers Close, was farmed by Mr Carter. It was called Boxworth End Close. Although it has probably been grassland since it was enclosed, from the ground and aerial photographs it is possible to make out the dozen or so strips which made up the field when it was still part of the open-field system.

By the time the first 25 inch (1:2,500) map was published in 1878 (Figure 4c) the building by the road (see page 21) had disappeared and the land beneath and beside it had been turned into part of the field which at that time lay between the road and Public Drain Number 12. Also, the hedge between Braziers Close and Boxworth End Close had been removed, turning them into one field. The site of the hedge is just discernible in the aerial photograph taken in 1988 (Plate 1). The big elm trees on the southern border of the field, which were a prominent feature when we bought it, are clearly shown in the 1878 map, as well as the larger elms on its western and eastern edges. In fact the field in 1878 was essentially the same as the one we bought in 1961 (Figure 4d). Changes brought about by me since I bought the field are also shown (Figures 4e and 4f) and are described in detail later (see Chapters 4, 6 and 7).

Thus, old maps show that our field was made up of a house or farm building, an enclosed close, and part of the original unenclosed Swavesey field system. The field's evolution since the 1830s can be traced in Figures 4a to 4f.

# CHAPTER FOUR

## EARLY DAYS

THE 1960s was the decade in which conservation became public property, and those of us earning our living in conservation research were kept extremely busy both doing research and trying to see that it was applied. Building up our research programme at Monks Wood on the effects of the persistent organochlorine insecticides on wildlife, and seeking restrictions on the use of these highly damaging chemicals, was a full-time job and I had little time to spend on our newly acquired field. I could deal with practical problems only as they arose and attempt in limited ways to make a small part of the field better for wildlife. However, these attempts provided a valuable reconnaissance for the future. In this chapter I describe them briefly as the setting for the main work which we undertook in the 1970s and 1980s.

The first practical problem was provided by the wall which had enabled us to buy the Farm House but which was in a perilous condition. The ditch, Public Drain Number 12, was undermining its footing, and it was obvious that most of the wall could not be made safe until I could move the ditch away from the edge of the wall into the field to its north (see Figure 1). This did not seem to be an insuperable task, but clearly I had to get the permission of the local Council to do it. The ditch was after all a Public Drain and the Council maintained it. Throughout 1961 – it was the year Yuri Gagarin became the first man in space – I sought permission and also complained to the Council about the pollution of the ditch. At last in 1962 I was told categorically that I could not move the ditch because it was 'Prescribed'. I had to admit defeat and live with the wall increasingly likely to fall into the ditch. In the years that followed I took down bits of it when they looked dangerous. Sometimes I was helped by nature, as when the great gale of January 3rd 1976 blew down a considerable length of the wall into the ditch. The last bit to fall was in 1993. Today only a stretch of 8 yards of wall remains. By widening the ditch at its foot and repairing the brick work it should last some time. It provides shelter for a garden seat and a sunning spot for red admiral and small tortoiseshell butterflies, common darter dragonflies and occasionally a great red underwing moth.

Mr Northrop continued to graze his cows in our field. Each day he took them across the road from his farm and brought them back for milking. Today this would cause a serious accident within minutes, but in the early sixties children played in the road and sheep were driven along it: there was no problem. Our younger daughter, Helena, armed with a little stick, often used to 'help' Mr Northrop drive Buttercup, Daisy and Twist from our field, through the gate in the hedge and then across the road to his farm. So far as I

Figure 5: The south-west corner of our field in 1965 before any planting had been done

Gypsy is standing on the site of what was to be the Old Plantation. Across Public Drain Number 12 lies the area later covered by the Buckingway Plantation, the Boundary Shelter Belt, the line of fruit trees, the Pond and the Winter Mere (Figure 6).

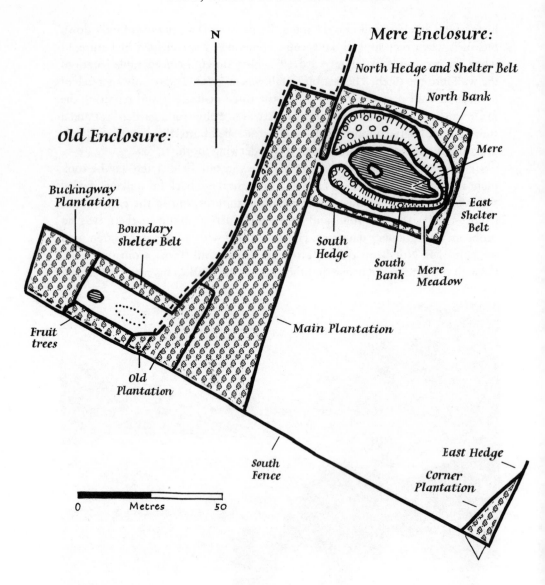

Figure 6: Sketch map of our nature reserve and corner plantation, 1998

Habitats: woodland, bramble bushes (circles), banks, water (hatched), Public Drain Number 12 (dashes) and site of the Winter Mere (dotted line) are indicated.

The Old Enclosure includes the Buckingway Plantation, the Boundary Shelter Belt, the line of fruit trees, the pond and the Winter Mere and the open ground surrounding them, and the small part of the Old Plantation north of Public Drain Number 12. The Mere Enclosure includes the Mere and its surrounding grassland, banks, shelter belts and hedges.

remember he never had more than five cows in our field. That was good for its few cowslips and very good for butterflies like meadow browns, and for thistles.

Grazing pressure was increased when our elder daughter Caroline's pony, Gypsy, arrived in 1964 (Figure 5). That year Mr Northrop retired and I let the grazing to Mr J. W. Burgess whose family farmed the land to the south and east of our field. The new grazing regime was much more intensive: initially Mr Burgess' herd consisted of about 50 Friesians and a Hereford bull. The latter was a docile beast with one eye: one could get very close to him without his noticing. The alteration in grazing regime produced changes in the butterfly population. Counts made on comparable transects in 1964 and 1965 produced 11 meadow browns and 4 large skippers in 1964 when the grass was long, but only 1 meadow brown and 1 small skipper in 1965 when it was short.

Mr Burgess took his cows into our field through a gate in the south fence, which connected the long line of elm trees mentioned on page 25 (see also Figure 1). We could get into our field from our garden so there was no longer a need for us to have a gate into the road; I took it down and planted the gap with hawthorns. The problem then was how to get Gypsy from her loose box in our yard into the field now that she could no longer go by the road. It was solved by Mr Burgess kindly allowing me to take her through a corner of his land adjoining our little orchard. I reinstated the gate between Mr Burgess' land and our field. Gypsy seemed happy to share the field with the cows, though she had a strong will of her own and liked to tease and exercise us by being difficult to catch. Eventually she became too small for Caroline and in 1966 she was sold. Her place was taken by a gelding, inappropriately named Abraham. However, in 1968 I found myself in the traditional role of horse dealer, extolling Abraham's virtues and glossing over his vices. I was successful and Abraham left for a new home.

In 1966 the Electricity Authorities had put a power line across our field. This was sad as it was unsightly and two fine elms in the south fence had to be felled to make way for it. In putting up the poles the contractors made severe ruts in the field. However they paid compensation for this and to this day I get a small rent for having two poles on my land.

Once the piece of the field by the road was no longer required as a drove for cattle or ponies, it was an easy matter to fence it off from the main part and turn it into something else. We called it the Old Enclosure (Figure 6). It was an odd bit of land – the site of an old building (see page 21) – containing much brick rubble, old and new, and part of it flooded every winter but then dried out completely in the summer. I decided to plant most of this bit of land, as well as about 420 square yards of the main part of the field adjoining it. To keep out the cattle from this extra bit I used an electric fence, but later replaced it with a stake and wire fence. My aim was to plant two little woods at either end of the enclosed land and to join them by a line of trees – the Boundary

Shelter Belt – and a line of fruit trees along the well-drained, rubble-filled edge of the ditch. To make access easier from our garden I replaced the somewhat hazardous plank bridge across the ditch by two much more solid bridges made of railway sleepers, which had been given me by my old friend and colleague, the late Gordon Mason, Warden of the Woodwalton Fen NNR. Alan Frost, a colleague at Monks Wood, kindly helped me set them up.

Between 1965 and 1970 I planted oaks, ashes, hazels, field maples, beeches, Scots pines, hollies, yews, a crab apple and a service-tree, and privet bushes and brambles in the little wood by the road, which I called the Buckingway Plantation (see Figure 6). In the little wood planted in the area fenced off from the main body of the field, which I called the Old Plantation (see Figure 6), I planted oaks, ashes, field maples, hornbeams, alders, willows and another service-tree. In between the two little woods I planted the Boundary Shelter Belt with oaks, alders, willows, wild cherry, leylandii cypresses and Lombardy poplars and, on the ditch-bank, six apples and a pear.

The trees came from many sources. I bought some from local nurseries, others were given by friends and relations. The service-trees were rescued as small seedlings from a ride in Monks Wood NNR which was about to be cut. Later, birches were rescued from a similar fate at Holme Fen NNR. Apart from those which had sown themselves in our roadside hedge, the ashes were seedlings which I had dug up in our garden. Willows came from cuttings from trees and bushes growing in the fens. Basically I planted what I could get. I would learn from what grew and what did not.

In general, oaks, ashes, field maples, hazels, willows and Lombardy poplars did well. Apart from one tree planted in a permanently damp spot, the alders did poorly. There were many casualties among the beeches, hollies and yews. All but one of the fruit trees did well. By the early 1980s the Buckingway and Old Plantations would look like proper woods (Plate 2).

Two of the Lombardy poplars grew remarkably well (Plate 3) and so did a big crack-willow. These trees would grow too well: the tallest Lombardy poplar grew about 30 feet in 24 years. In 1991 I had to have them felled and some oaks coppiced because they would pose a potential threat to our neighbours if we were to get another gale of the ferocity of those which occurred in 1976 and 1990. Our neighbours were also concerned about subsidence which they thought might occur as the result of tree roots drying out the clay soil.

Back in the 1960s, our garden lacked that most essential feature, a pond. All we had was an old tin bath rescued from a rubbish dump. It attracted a great crested newt but otherwise had little to be said for it. So, in 1968 I dug a pond in the low ground between the two little woods (see Figure 6) and lined it with a sheet of butyl 15 feet long and 12 feet wide. I filled it with tap water and planted two water lilies in it. It was to hold water for over twenty years but eventually roots from a neighbouring willow tree pierced the butyl and it dried out in the summer months (shown on Plate 32).

To the east of the pond was the quite large area of slightly lower ground which was wet in winter and dry in summer. It presented a challenge because it was too wet to plant and yet could not be made a permanent pond without lining it, and that would have been prohibitively expensive. We had to accept it for what it was and called it the Winter Mere (see Figure 6). We could at least make it live up to its name, so in 1970 I dammed its eastern end and enlarged it. By the winter it was just large enough to skate on and we did. In summer we used it as a place on which to burn rubbish.

The effects of turning c.820 square yards of field into two little woods joined by a shelter belt and a line of fruit trees, a pond and the Winter Mere were interesting and gave useful pointers for the future.

The two small woods were used by bird species which already occurred in our garden, such as blackbird, dunnock and wren, but no new species colonized them. However, insects which I had not seen before did appear, for example the beautiful eyed hawk-moth and the puss moth, which has that weirdest of caterpillars. The larvae of both these species must have fed on the willows and poplars which we had provided.

Smooth and great crested newts soon discovered the pond, and the smooth newt certainly bred in it. The pond was also colonized by blue-tailed and azure damselflies. In the very cold winter of 1962–63 I picked up a snipe by the ditch. It was clearly in a bad way so I took it indoors to warm it up. In a surprisingly short time it began to fly round and round our dining room. It was a most weird sight. I released it and I hope it survived. On another occasion I put up a grey wagtail from the ditch. It must have found the polluted water a poor substitute for the streams of the west. Neither the pond nor the Winter Mere attracted any water birds.

Much the most obvious increase of wildlife was in the butterflies in the remaining grassland. When counts on 310 yards of transect in the intensively grazed, unenclosed part of the field were made on 18th July 1965 (see page 63) only two butterflies – a meadow brown (Plate 4) and a small skipper – were seen, yet counts on only 110 yards of transect in the enclosed part of the field made on the same day revealed 22 butterflies – a small tortoiseshell, 15 meadow browns, 4 ringlets (Plate 5), and a large and small skipper.

I made numerous counts in the enclosed area in July and August in 1970 and 1971. They showed interesting differences from the counts made in the sixties. The ringlet remained, but the meadow brown had disappeared and its place had been taken by the gatekeeper (Plate 6) which was now much the commonest species in the Old Enclosure. Meadow brown, ringlet and gate-keeper caterpillars all feed on grasses. I noticed that the flowers of the bramble bush I had planted (see Plate 3) were much favoured by adult gatekeepers.

My activities in this reconnaissance period of 1960–70 had shown me which tree species grew best on the heavy Ampthill Clay soils. Not surprisingly they were those characteristic of ancient woodland on the clays of Cambridgeshire – oak, ash, hazel and willow. It looked as if the wood must be made larger if it were to attract bird species which were not already present in our garden. The small pond surrounded by trees was quickly colonized by newts but very few dragonflies colonized it. On the other hand butterflies had increased rapidly when long grass and shelter were provided. Nevertheless my records showed that butterfly populations could change within a short time. When opportunities arose I could put what I had learnt into practice. As things turned out, after 1970 I was stimulated to act quickly by an event which had a devastating effect on the Cambridgeshire landscape.

Figure 7: The north-west part of our field in 1965, showing two large elms growing on the edge of Public Drain Number 12 and small ones in the overgrown hedge to the north. Gypsy is standing on the site of what was to be the Main Plantation. Directly behind her is the finest of the elms which grew round our field.

# CHAPTER FIVE

# DUTCH ELM DISEASE: A SPUR TO ACTION

THE most striking and delightful features of our field were the elm trees in the hedges and fences which surrounded it. Only those on the eastern hedge were mine but all were enjoyed. The most shapely was the huge tree with its feet in the ditch on the western border: it had a good water supply and room to expand (Figure 7). The southern border consisted of a line of 24 elms – those shown on the 1878 25-inch map (see page 25). The tallest could be seen for miles around. Coming home from work it was good to see them from the Roman road (now the A14). When, sadly, I had the opportunity to count their growth rings I found they were of different ages. The oldest which I could study had a diameter of three feet and was planted about 1848 at the time of the enclosures. Several of the trees were pollards and most of these had been allowed to grow up. Most of these elms were showing signs of age. The elms in the eastern hedge (Plate 7) were much younger: the tallest one was only c.50 feet high. One of 35 feet which I could age was planted or appeared in 1942, a neighbouring one in 1935. It would be some time before these could rival the giants in the southern fence but they showed promise. There were about a dozen sizeable trees mainly in the southern part of the hedge; the rest were little more than suckers. The northern boundary of the field was a thick hawthorn hedge but it had two elms on its eastern edge.

Dutch Elm Disease, which is caused by the fungus *Ceratocystis ulmi* and is spread by *Scolytus* bark beetles, arrived in imported timber in 1969 and spread rapidly. At first there was an attempt to stop its spread by felling elms in its path. This was soon shown to be ineffective and 'sanitation felling' was halted in 1972. We dreaded the spread of Dutch Elm Disease to Cambridgeshire particularly, because apart from large pollarded willows in the fens, nearly all the large hedgerow trees in the county were elms. Hedgerow oaks grew slowly and were rare, and while ash quickly colonizes Cambridgeshire hedges it rarely does well as a mature tree. Apparently only the elm could live happily in arable areas with a low rainfall. If we were to lose the elms the Cambridgeshire landscape would become virtually devoid of large trees. Sadly this is what happened.

In 1966 two old elms had to be felled to allow the power line to cross our field, but apart from that we lost no elms from 1960 to 1972. Then that year Dutch Elm Disease struck. I tried to slow the advance by burning affected branches or whole trees in the east hedge (see Plate 7) but to no avail. Eventually in 1979 I had to fell them all bar one small one which I left as a memento but which died soon after (Plate 8). In 1975 the splendid great elm on the western edge had succumbed. On 3rd January 1976 the ferocious gale felled one stricken tree on the southern fence (Plate 9). All the other great elms on that fence were becoming a hazard to man and beast (Plate 10) and in 1979 my neighbour (their owner) had to fell them too. At my request he left some pollarded stumps. Stock doves and a little owl continued to breed in them for a while but then they disappeared and today the only signs that our field was once surrounded by magnificent elms are slight depressions in the ground, and three or four elm saplings with some suckers in the east hedge.

As for many others, the loss of elms was a spur to tree planting. I would have probably planted more trees in any case but the loss of our elms made me get down to it earlier than I would have otherwise done.

# CHAPTER SIX

# PLANNING AND PLANTING THE
# MAIN WOOD

I had always hoped that I would be able to plant more woodland in our field. Now, stimulated by the loss of the elms, I began to plan in earnest. There is something awesome about a blank sheet of paper, but once one begins to sketch and doodle the ideas begin to flow. Of course the first thing I had to ask myself in 1970 was what am I aiming to achieve?

All around us woods were being destroyed: 17 per cent of the area of woodland in western Cambridgeshire had been lost since the end of the Second World War. Although I could not create an ancient woodland I could at least make a plantation of native trees which would support some of the plant and animal species found in ancient woodland. But should I plant several small woods or one big one? From a landscape point of view several small woods might look nice. However, experience with the Buckingway and Old Plantations and my own studies of bird populations of woods of different sizes had shown me that if I wanted more bird species the wood must be as large as possible. Therefore I must have one large wood, but how large? My time was very limited so there was a limit to the amount of planting and managing I could do. I did not want to apply for any of the grants available for planting woods under the Forestry Commission's excellent scheme, because I wanted complete control over what I planted and where. I had not got the resources to plant the whole eight acres, even if I wanted to, and I did not: I wanted a mixture of habitats. The wood had to be as large as possible, but manageable.

Where was the wood to be? Several considerations came into play. Fencing is expensive so I needed to have as little fencing as possible. I might need to water young trees in their first years of growth so the wood must not be too far from a source of water. Thirdly, I could make a larger wood if the Main Plantation which I planned could be joined to the plantations I had already planted. All these points would be catered for if I planted the new main plantation on the west side of the field (see Figure 6). The ditch required no fencing so I need not fence the long west side of the plantation nor its short northern boundary. The plantation would be as near to our garden tap as it could be and it would join up with the Old Plantation, the Boundary Shelter Belt and the Buckingway Plantation, thus making a woodland of about one acre overall. If I wanted to add to this I could always do so if opportunities arose. In 1971 I marked out the fence line of what we were to call the Main Plantation. I would need a fence of nearly 140 yards.

Having decided where to plant I had to decide on what to plant. Experience with the Buckingway and Old Plantations had shown that the tree species which did best were those which thrived in ancient woods on the Cambridgeshire clays and thus probably existed on the site of my field before it was cleared for agriculture. Therefore I must concentrate on oak, ash, field maple, hazel and willow and retain any elm that happened to survive as suckers from the field boundaries. It would be fun to experiment with other rarer species which might have grown there or nearby in the distant past. I would plant some alders, aspens, grey poplars, birches, wild cherries, horn-beams and even try out some which probably had not been there, such as beech, small-leaved lime, gean and rowan. I also wanted trees that would grow quickly so that I would have trees to look at within a few years rather than small saplings: I decided to plant quick-growing hybrid black-poplars and Lombardy poplars by the ditch. Finally I wanted some evergreens both to provide roosting places and cover for small birds and as a shelter and screen on the southern border of the wood. Yews and hollies, native to Britain but almost certainly not on the Cambridgeshire clays, had grown very slowly or had died in the Buckingway Plantation so were clearly not suitable. So I tried Scots pine and, when they showed signs of doing poorly, I planted *Cupressocyparis leylandii* which I knew would grow quickly almost anywhere. I wanted to turn the fence into a hawthorn hedge; some of these hawthorns might later be allowed to grow into trees to provide food for wintering redwings and fieldfares.

Young trees are extremely expensive if bought singly or in small numbers from garden centres, but they are much cheaper if obtained in bulk from a nursery. It is generally best to plant young trees fairly close together so that they shade each other and competition makes them grow taller as they seek the light. I reckoned that the area to be planted was about three-quarters of an acre and therefore at a spacing of two yards between trees I would need about 360 trees. That was a lot and even if bought from a nursery would be costly. By extraordinary good fortune I learnt from my colleague, Dick Steele, who at that time was in charge of the Woodlands Section at Monks Wood, that about 400 young oaks grown at the Nature Conservancy's nursery at Rostherne Mere National Nature Reserve in Cheshire were surplus. 'Would you like them for your wood?' Indeed I would. It was a chance not to be missed. It would save me a considerable amount of money and make use of a lot of trees which otherwise would probably have been destroyed. As mentioned above, I had originally planned a mixed wood of at least five species; I now had to alter the plan by which I could achieve this aim. I decided to plant all the 372 oaks which had arrived from Cheshire. Some of them were bound to die; I would then plant the gaps left by the dead ones with the other species as opportunities occurred. Thus, starting with a matrix of oak trees I would end up with a mixed wood, though oak would probably be the most abundant species at the end of the day.

Mr Burgess kept no cattle in our field during the winter of 1971–72 so there was nothing to prevent our starting to plant the oaks. I say 'us' because I got considerable help from my son, Peter, who was an undergraduate at Cambridge at the time, and from our friend Robin Buxton. We started to plant in December 1971, putting in 34 trees near the markers of the fence line. It was hard work digging up the waterlogged turf and then digging deep enough down so that the roots could be spread out. The holes often filled with water and the heavy clay soil was like wet plasticine and difficult to break up when filling the holes after the trees had been planted. We continued to plant through the Christmas holidays and for three weekends after that (Plate 11). Our younger daughter, Helena, planted the last tree on 23rd January 1972. We had planted 372 trees in ten days – 170 in December and 202 in January. The most we planted in one day was 97 on 16th January 1972. On 19th January I stuck in 24 willow cuttings near the ditch, and on 27th February I planted 15 hybrid black-poplars in a line by the ditch. Mr Mitham, the local builder and our County Councillor, put up the post and wire fence in early April.

On 5th July 1972 we left for Australia for two months where I was to advise CSIRO (Commonwealth Scientific and Industrial Research Organisation) about the effects of pesticides on waterfowl which were dependent on irrigation systems in the outback. This meant that for two important months the little trees in the Swavesey Wood would not get watered. The situation would be serious if there was a drought. There was: 148 (40 per cent) of the oaks were dead by the end of the year. There would be plenty of spaces for species other than oak!

In 1973 I began to make the Main Plantation more varied (Plate 12). I planted the Scots pines on the southern edge as mentioned above. I planted seven beeches, ten hornbeams, two ashes and two walnuts. By 1975 I had planted 482 trees altogether: 459 belonging to seven native species and 23 to

Table 1    Survival of native trees planted in the Main Plantation, 1971–75, to the end of 1975

| Species | Year Planted | Number planted | Number survived |
|---|---|---|---|
| Oak | 1971/2 | 372 | 171 |
| Silver Birch | 1975 | 54 | 3 |
| Ash | 1973/4 | 13 | 11 |
| Hornbeam | 1973 | 12 | 9 |
| Beech | 1973 | 7 | 4 |
| Small-leaved Lime | 1973 | 1 | 1 |

Note: Hazels and Field Maples planted after 1975 did well as did Field Maples in the Old Plantation.

exotic species. The survival rates of the native trees by 31st December 1975 are shown in Table 1. All the non-native species (15 Lombardy poplars, two cherry plums, two sycamore, one holm oak, one walnut, one ginkgo and one metasequoia) all survived except for four poplars, one cherry plum and the ginkgo. I continued to plant in the Main Plantation up to 1982. Hazels and field maples were particularly successful. Among the trees given me were two ashes provided by the Parish and five small-leaved limes given me by the late Jim Grant, who was studying the biology of the little native cicadas which fed upon them. I also planted a yew in 1976; 20 Lombardy poplars, two walnuts and two ashes in 1977; and three oaks, transferred from the Old Plantation, in 1978. In 1980, I transplanted from the garden a horse chestnut which Peter had grown from a conker when he was a small boy. In 1981 three more Lombardy poplars went in and, in 1982, ten field maples, ten geans, ten oaks and ten white poplars were planted. That year I also planted a compact group of leylandii cypresses at the northern end of the wood.

After 1982 the trees in all three plantations were left to fight it out with little interference or support from me. An inventory of the trees surviving in 1997 is shown in Table 2. By 2000 the situation had changed little except that five more silver birches and the metasequoia had died. Losses appeared to be due to a combination of factors. Fortunately there were no rabbits and hares were sadly rare. Field and bank voles were common and in some years did do some damage to young trees. I often had to wrap plastic round the lower parts of trees to protect them from voles. This treatment seemed to be effective though one year two quite large trees were attacked below ground level and died. In general wild mammals had little effect. Lack of water was quite another matter. Apart from the loss of 40 per cent of the original oak trees in 1972, the severe droughts of 1991, 1992, 1994, 1996 and 1997 took their toll: many

Table 2    Inventory for 1997 of all trees planted 1965–85, and their survival rates

**Number surviving in the plantations indicated**

| Species | Buckingway Plantation | Old Plantation | Main Plantation | Corner Plantation | Mere Enclosure | Total Number Planted | Total Number alive 1997 |
|---|---|---|---|---|---|---|---|
| Pedunculate Oak | 19 | 13 | 143 | 14 | 16 | 436 | 205 |
| Ash | 9 | 28 | 24 | 6 | 24 | 100 | 91 |
| Silver Birch | 3 | | 10 | | 3 | 62 | 16 |
| Hazel | 7 | | 23 | | | 60 | 30 |
| Willow spp.[1] | 6 | 4 | 4 | | 10 | 60 | 24 |
| Elder | 6 | | 1 | | 40 | 55 | 47 |
| Field Maple | 7 | 11 | 7 | 17 | | 50 | 42 |
| Cupressocyparis leylandii | 8 | | 22 | | | 40 | 30 |
| Hybrid Black-poplar | | | 11 | | 12 | 35 | 23 |
| Hornbeam | | 6 | 6 | | 5 | 27 | 17 |
| Lombardy Poplar | 3 | | 16 | 1 | 2 | 23 | 22 |
| Yew | 5 | | 1 | | 10 | 20 | 16 |
| Elm[2] | | 7 | 6 | 3 | 1 | | 17 |
| Alder | 4 | 1 | | | | 15 | 5 |
| White Poplar | | | 4 | 6 | | 15 | 10 |
| Apple | 10 | 2 | | | | 13 | 12 |
| Beech | 6 | | 2 | | | 15 | 8 |
| Gean | | | 2 | | | 10 | 2 |
| Holm Oak | | | 4 | | | 10 | 4 |
| Scots Pine | | | 4 | | | 10 | 4 |
| Small-leaved Lime | | 2 | 6 | | 1 | 10 | 9 |

Note:
In addition the following species were planted in small numbers (their survival rates are shown):
Holly 2/5, Firethorn 5/5, Privet 5/5, Service-tree 1/3, Walnut 2/3, Sweet Chestnut 1/2, Buckthorn 1/2, Indian Horse-chestnut 2/2, Sycamore 2/2, Sessile Oak 1/1, Black-poplar 1/1, Crab Apple 1/1, Pear 1/1, Horse-chestnut 1/1.
15 Red Alder, 4 Larch, 2 Aspen, 1 Spruce, 1 Rowan, 1 Alder Buckthorn, and 1 Ginkgo were planted but none survived.

[1] Three species, but mostly Crack-willow. All were planted as cuttings.
[2] All but the one in the Mere Enclosure were naturally grown suckers.

birches died and so did the tops of several ash trees. Undoubtedly drought was the main problem we faced in establishing the woods. I did some watering but I could never provide enough. I tried to reduce water-loss by weeding the bare ground round the young trees thus reducing competition from grasses. At a later stage, competition for water and light between trees must have exacerbated the effects of drought.

Despite the setbacks the wood developed satisfactorily. It was a huge pleasure to see a grazed field turn into rough grassland with little trees emerging above the grass and thistles, and then in turn to see it all turn into uneven oak scrub (Plates 11–17). About 1986 most of the ground became shaded by the trees and the whole area began to look like the wood it would become in 1990 (Plate 18). At this time the ground flora changed completely from one dominated by grasses to one consisting almost entirely of cow parsley with a little ivy and ground-ivy in places. It looked just like the ground flora of other plantations which I had seen in Cambridgeshire when studying their birds. Some of these plantations were over 100 years old so it was clear that if I wanted something more interesting in the ground flora I must introduce the species I wanted to see.

One of the botanical specialities of south-west Cambridgeshire is the oxlip, which replaces the primrose in those ancient woodlands which grow on chalky boulder clay. Since my wood was not on chalky boulder clay I could not hope to establish the oxlip. However primroses do occur in Cambridgeshire, and the cowslip, which is quite common in unimproved meadow, road verges and on the sides of ditches in the county, also occurs in the rides of woods. So I decided to try and establish primroses and cowslips, together with bluebells, wood anemones and dog's mercury, which also occur locally in ancient woods. When a friend in the Lake District offered me a few wild daffodils I accepted these too. This species is probably native to Cambridgeshire though now very rare. I also planted two summer snowflakes in a damp spot; they were given me by another friend. It is a rare native species mainly found in the Thames catchment area.

The reason that the delightful flowers of ancient woodland do not occur in plantations dominated by cow parsley might be due to the nature of the soils of plantations or to the competition they face from cow parsley. In 1975 I started an experiment to determine the answer. I planted bluebells in four separate plots having weeded out the cow parsley. In two of the plots I did no further weeding, allowing the cow parsley to recolonize them. In the other two I weeded out cow parsley seedlings each year. After seven years there were no bluebells in the plots where I had done no weeding, but in the two weeded plots the bluebells survived. It suggested that bluebells could not compete with cow parsley; the reasons can only be guessed at. Cow parsley has long carrot-like roots and appears to absorb much moisture from the soil; perhaps it makes the soil too dry for bluebells and other species to co-exist with it. Also, later in the year when cow parsley grows thick and tall it must prevent most of the light from reaching the bluebells beneath it.

By 1977 I was sufficiently sure about what the results of this experiment would be to start planting woodland plants in other places, weeding them where necessary. From that year onwards I planted them at irregular intervals. I have recorded their presence and absence every year since, and in most years between 1977 and 1992 I recorded the actual number of plants, the number of flowers and the approximate number of leaves. Since the plantings were done at different dates the situation is complicated but what happened can be summarized as follows.

Bluebells survived where planted and sometimes produced offspring, though these small plants often disappeared in subsequent years. Bluebells did better in the slightly damper plots. The number of flowers produced varied from year to year. For example, in the 1982–92 period it varied from 43 to 112 in one plot, 17 to 60 in another and 0 to 12 in another. The highest numbers in each plot did not coincide. Leaf numbers also varied: some plants gradually increased their number of leaves, in others it declined. All this suggested that bluebells were reacting to small differences in soil, moisture and shade.

Wood anemones also survived where cow parsley was weeded except in one rather damp plot. However, plants were usually small, they produced few flowers and never looked really happy.

Dog's mercury produced flowers but never became luxuriant, as it does in many Cambridgeshire woods. Nevertheless each clump spread quite widely.

Of the ten primroses planted in the wood only four survived to 1997. This was in contrast to our orchard where five out of six survived.

Of the 12 cowslips planted in the Swavesey Wood in and before 1983 only three clumps remained in 1996. Interestingly this was a better survival rate than in the Mere Meadow (see page 52) where all six planted in 1984 had died by 1995. The surviving clumps consist not only of old plants but of seedlings which took root beside them. They are weeded annually and still look healthy in 2002 (Plate 19).

The wild daffodils have done well (Plate 20). Not only have the plants survived and the bulbs subdivided, but the flowers produced have greatly increased. From 1980–85 there were never more than four flowers in any one year, but since then they often produce many more: in 1996 there were 66 flowers.

The snowflakes have soldiered on but they look less happy than plants of the same species which grow under more open conditions in our garden.

Since 1992 I have done less weeding of cow parsley so bluebells, anemones and dog's mercury have declined although this may be due partly to the droughts experienced in 1994 and 1996. In the very dry year of 1997 flowering was suppressed: only 22 of 58 bluebell plants, two of six primroses, and three of six cowslips flowered. There were about 90 wood anemone leaves but only three flowers, and about 140 shoots of dog's mercury of which 32 had flowers.

The overall conclusion was that, while the common woodland wild flowers can be grown in a young plantation of deciduous trees, they require almost as much attention as if they were being grown in a garden. This was the result I had expected but it was good to provide yet more evidence that the ancient woods which survive in Britain are extremely precious because they cannot be re-created. However, to see woodland flowers appear each spring in the Swavesey Wood has been a continuing delight and well worth the laborious business of pulling out cow parsley roots.

I have to admit to being fond of cow parsley: when in full bloom it makes a splendid sight on the verges of roads and lanes in Cambridgeshire, and when it grows in the sun it attracts many insects. However, as a monoculture in the Swavesey Wood it was rather boring. Would or could any other plant compete successfully with it? The old elms had been ivy-clad and when I planted the wood ivy soon began to invade land near the fence. Where this happened the cow parsley disappeared. The only other plant which appeared to be able to compete with cow parsley was goose-grass or cleavers, the plant which children love to attach to the clothes of unsuspecting grown-ups, but which has little else to be said for it. A woodland flora of ivy and cleavers did not appear to be an improvement on cow parsley. Could I introduce something else? I had noticed that lesser periwinkle, an attractive plant with evergreen leaves and pale blue flowers, did well in some shady places in south Cambridgeshire. So in 1984 I planted a few plants in the Old Plantation. With minimal weeding these had formed a patch two yards across by 1989, and by 1997 the patch was four yards long and three yards across and threatened to overrun an area with bluebells and anemones. It remains to be seen if it can extinguish cow parsley completely without any help. Nearby I also planted the closely related greater periwinkle, a common garden plant. Initially it did badly, but when a nearby tree died and let in the light it began to do better and it now occupies an area which is about the same size as that occupied by the lesser periwinkle. Thus at present, the only species which can take over from the cow parsley are the native ivy and the two periwinkles, of which the greater is certainly not native, and the lesser is almost certainly not native in Cambridgeshire. It is interesting that all three species are evergreens and send runners along the soil surface.

As the years go by I expect ivy will eventually take over the floor of the wood, though when a tree dies or is blown over and the light is let in grasses, ground-ivy, docks, nettles and brambles may invade, as they already have in a few spots. In 1997 the area covered by ground-ivy in the wood had greatly increased so that over 50 square yards were covered by this species. I suspect that it has been able to increase because the drought has caused many tree

Table 3     Regeneration: seedlings and young trees in the Swavesey woods and grasslands, September 1997

**Number of seedlings and young trees in the localities indicated**

| Species | Buckingway Plantation* | Old Plantation | Main Plantation | Mere shelter-belts | Mere grasslands and banks | Old Enclosure grassland |
|---|---|---|---|---|---|---|
| Field Maple | | 6 | 122 | | | |
| Sycamore | | | 81 | | 3 | |
| Ash | 1 | | 41 | 6 | 2 | 2 |
| Hawthorn | | | 12 | 11 | 8 | 1 |
| Elder | 3 | | 3 | 9 | 3 | |
| Oak | 2 | 1 | 13 | 1 | 1 | |
| Hazel | 1 | 1 | 15 | | | |
| Gean | | | 2 | | | |
| Yew | 1 | | | | | |

Notes: (i)   * Includes Boundary Shelter Belt and area under fruit.
      (ii)   No seedlings or young trees were found in the Corner Plantation.

branches to die and this has let in more light.

Over the years I have often observed seedlings of different tree species in the woods and grasslands. In the late summer of 1997 I made a careful count of all that were visible at that time (see Table 3). Seedlings of nine species were recorded. Most were in the Main Plantation, but their distribution there was very uneven. In the part of the wood which is crossed by the path, 166 out of 280 seedlings (57 per cent) were found within one yard of the edge of the path – that is very approximately in one-eighth of the area. Most seedlings of maple, oak, hazel and hawthorn were by the path, while most of the ash and sycamore ones were away from it. As the path receives slightly more light than other areas this suggests that ash and sycamore are more tolerant of shade than the other species. It is interesting to note the success of the sycamore: only two sycamores have been planted, yet over 25 per cent of the seedlings in the wood were sycamores. Whatever happens to the ground flora it seems likely that sycamore, field maple and ash will become increasingly important trees in the wood if any areas are opened up.

Many seedlings were found in 1997 but it is obvious from the scarcity of larger ones (see Table 4) that their growth has been slow: only six were more than three feet tall. By 2000 there were 16 seedlings over three feet tall (13 maples and one each of ash, sycamore and yew). The tallest was a ten-foot high maple. The unusually wet summer that year probably encouraged their

Table 4    Heights of seedlings and young trees in Swavesey woods and grasslands, September 1997

**Heights in Woodland**

| Species | <1 <0.3 | 1 0.3 | 2 0.6 | 3 0.9 | 4 1.2 | 5 1.5 | 6+ feet 1.8 metres |
|---|---|---|---|---|---|---|---|
| Field Maple | 100 | 25 | 1 | 1 | 1 | | |
| Sycamore | 68 | 10 | 2 | | 1 | | |
| Ash | 40 | 7 | 1 | | | | |
| Hawthorn | 21 | 3 | | | | | |
| Elder | 6 | 6 | 1 | 2 | | | |
| Oak | 16 | 1 | | | | | |
| Hazel | 15 | 2 | | | | | |
| Gean | 2 | | | | | | |
| Yew | | | 1 | | | | |

**Heights in Grassland**

| Species | <1 <0.3 | 1 0.3 | 2 0.6 | 3 0.9 | 4 1.2 | 5 1.5 | 6+ feet 1.8 metres |
|---|---|---|---|---|---|---|---|
| Field Maple | | | | | | | |
| Sycamore | | | | | | | 3 |
| Ash | | | 1 | | | 1 | 2 |
| Hawthorn | | 1 | 1 | 2 | | 2 | 2 |
| Elder | | | | 1 | | | 2 |
| Oak | | 1 | | | | | |
| Hazel | | | | | | | |
| Gean | | | | | | | |
| Yew | | | | | | | |

Notes: (i) Woodland includes Boundary Shelter Belt and area under fruit.
(ii) Grassland includes remaining grassland in the Old Enclosure as well as the Mere meadow and banks.

growth. The situation in the Mere grasslands (see page 52) was noticeably different from that in the wood. In 1997 there were fewer seedlings, presumably because the grassland had been enclosed for only 13 years compared to the wood (26 years). On the other hand a much larger proportion had developed into saplings, presumably because they had received much more sunlight than the seedlings in the wood. It is interesting that after 19 years no seedlings have developed in the Corner Plantation (see below), which is a drier more exposed wood than the Main Plantation.

When John Burgess cut down the dead elms along the south fence the field looked terribly bare, so I decided to establish another plantation mainly for landscape reasons rather than for wildlife. In 1981 I discussed the matter with John and we agreed to make a corner plantation in the south-east corner of my field, extending it a little to the south on his land. He kindly undertook to do the fencing and I did the planting. I planted 20 oaks, ten field maples, ten white poplars and one Lombardy poplar and protected them with a hawthorn hedge made up of a hundred plants on the western and southern sides. As with the main Swavesey Wood there were early casualties and I replaced them with trees not necessarily of the same species as those which had died. In 2000 the Corner Plantation (Plate 21), which covers c.270 square yards, contained 18 field maples, 14 oaks, six ash, six white poplars, a Lombardy poplar and four tall elm suckers. The hawthorn hedges have grown well. The ground flora, like that of the Swavesey Wood, consists mainly of cow parsley with some cleavers.

A little before planting the Corner Plantation I had planted five oaks as hedgerow trees in the eastern hedge to replace the elms which I had lost. Not surprisingly they grew slowly, having to compete with the roots of well-established hawthorns (see Plate 21). Also they suffered from browsing by cattle despite wire fencing put up to protect them: it is easy to underestimate the length of a cow's neck! However, all five oaks survived. Two of them later received extra protection by being included in the eastern border of the Corner Plantation. These shaded and protected trees are now nearly twice as tall as the three exposed trees in the hedge outside the Corner Plantation.

By 1983, the year I retired from working for the Nature Conservancy Council, I had a well-established wood of about an acre and a young but viable Corner Plantation of about 270 square yards. I could now think about creating other habitats. This would involve planting more trees, but for quite a different reason, as I shall explain in the next chapter.

# CHAPTER SEVEN

## PLANNING AND PLANTING THE MERE
## AND ITS SURROUNDINGS

I N theory retirement should provide more time for leisure, but as many
discover it does not turn out quite like that. However, I was determined
to do what I had always wanted to do – construct a pond purpose-built
for dragonflies.

Few groups of animals have given me more pleasure than dragonflies. I can
remember when I first really looked at one. I had caught a common darter and
it nipped my finger. I marvelled at its intricacy. Another boyhood memory was
of blue-winged beautiful demoiselles fluttering energetically over a New Forest
stream. I did not know the names of either of these dragonflies: there was no
popular book about them. When so much was written about butterflies this
seemed very strange. Inability to name the dragonflies which I saw delayed my
studies of them. But in 1937 my problem was solved by the publication of *The
Dragonflies of the British Isles* by Cynthia Longfield, the Irish entomologist
and traveller. I was soon corresponding with her. Many years later she, our
entomological friend, Philip Corbet, and I together wrote a book entitled
*Dragonflies* for the New Naturalist Series.

What is so fascinating about dragonflies? I suspect that there are several
reasons. As insects go they are large and so one can see the details of their
remarkable construction. Unlike most insects, you can watch them in the field
with binoculars. In the book mentioned above I called them the 'bird watchers'
insect' – indeed they are. They are very beautiful: their eyes are large and
translucent with hidden depths. Several dragonflies including the huge

emperor, a common insect in southern England, and many damselflies are bright blue, which is a rare colour for animals to be. They seem to mirror the sky. Dragonflies are masters of the art of flying – darting forward at great speed, gliding, hovering and even flying backwards. The males of many species are territorial so they return to the same place again and again so one can have yet another look and then another. Dragonflies are essentially tropical animals: in Britain they are active only when it is warm. The larval stage is spent entirely in water, therefore the adults are mainly found by water. As a result of these requirements one associates dragonflies with pools, rivers and streams on fine sunny days: it is not surprising that people get hooked on dragonflies. And when one learns that they have been around for hundreds of millions of years, and have seen both the rise and the fall of the dinosaurs, one cannot but view them with admiration and awe.

Delight in dragonflies led to my doing research on their behaviour and ecology. Over the years I have been forced by circumstances to travel considerable distances in order to get from my home to my research sites. This meant that I could rarely study a population in one place day after day. How wonderful it would be to have dragonflies on one's doorstep where one could watch them whenever one wanted to. Once I retired in 1983 that seemed possible, and a very kind leaving present from my Nature Conservancy Council colleagues made it practicable. My colleagues had asked me what I wanted when I retired and I said a pond. With great generosity they gave me more than I needed to hire a JCB for a day and a half and dig a pond to my exact specifications.

Most dragonflies require permanent, unpolluted water in which their larvae can develop. The larvae need waterweeds from which to stalk their prey and in which to hide from predators. They also require tall emergent water plants up which to climb before turning into winged adult insects. Their habitat must be warm so the water in which they live should not be too deep nor too shaded by trees. The adults also need warmth, particularly sunny places out of the wind where they can catch their insect prey. All these points had to be taken into account when I planned the pond and its siting.

I decided to call the new pond 'The Mere' to distinguish it from the small butyl-lined pond which I had constructed between the Buckingway and Old Plantations. The word mere had good local connotations because all the extinct fenland lakes in the district were called meres: Whittlesey Mere, Ugg Mere, Brick Mere, Ramsey Mere, Willingham Mere, Stretham Mere and Soham Mere.

My observations over the years had shown me that large farm ponds generally supported more species of dragonflies than small garden ponds. It was not surprising that my little pond had supported only two or three species. Clearly the Mere must be quite large; there was not room for it in the remaining open ground between the Buckingway and Old Plantations. There-fore it had to be put in a new enclosure in the field. Where should it be dug? If I put it near the Main Plantation the latter would provide good shelter from the prevailing south-west wind, but I had to decide which end. If the Mere were dug at the southern end it would be nearer our house, but it would also be close to our neighbour's farm buildings which at that time supported a very large population of house sparrows. These birds are adept at catching emerging damselflies and could have a serious effect on their populations by the Mere. So I decided to site the Mere beside the north end of the Main Plantation (see Figure 6).

Out in the field the Mere would be exposed to winds from the north, east and south. It was particularly important to shelter it from the cold winds of the north and east. Protection from these could be achieved by making wind-breaks of the spoil dug from the site of the Mere, by planting a shelter belt on the northern and eastern sides and by planting a hedge round the whole enclosure (see Figure 6). No trees or bushes would be planted close to the Mere where they could eventually shade it. As a result the Mere would be surrounded by rough grassland; this would provide many small insects on which the dragonflies could feed. Indeed on the south side of the Mere, where the spoil bank would take up less space than on the north side, a little meadow would be left between the bank and the hedge (see Plate 30). Experience with the grassland between the Buckingway and Old Plantations suggested that it would provide an additional bonus: a good habitat for butterflies.

On 9th December 1983 we began work. John Burgess kindly allowed the contractors' JCB to go through his yard and field. I explained to the contractor what I wanted and where. The Mere was to be pear-shaped with its long axis running from east to west. There was to be an island towards the western end so that if a mallard chose to nest there she would be safe from foxes. The narrow east end was to be ditch-like and marshy. Most importantly the north side facing the sun must be shallow so that water there would be warm. The deepest part of the Mere to the south-west of the island would be about five feet in depth – deep enough to discourage its colonization by tall water plants which might eventually threaten to take over the Mere, and also to provide permanent water at times of drought.

The contractor took seven hours to dig out the pond on 9th December – each scoop taking out about a ton of clay (Plate 22). He returned on 12th December to finish off the work and tidy up. He was interested in natural history himself and sympathized with my objectives. He did a very good job. It cost £130. The great hole was surrounded by a narrow strip of the original grassland, outside which was an embankment made out of the clay spoil. On the northern side it varied from about four to six feet high (Plate 23), and was about three feet high on the southern and eastern sides. The turfs which he had removed before digging the hole were placed on top of the clay banks thus adding to the amount of grassland and preventing the clay from being washed away by rain. As soon as the contractor had left I staked out the fence line. Then John Burgess very kindly put a post and wire fence 170 yards long round the site (see Plate 23). Its western side was already fenced by the hedge of the Main Plantation. I had to make a gap in it so that I could get to the Mere from the wood.

Introducing water plants into the Mere would have to wait until it had filled up with water, so in January 1984 I turned my attention to the shelter of the Mere rather than the Mere itself. I began to plant the hedges on the circumference and the trees for a shelter belt on the north (see Plate 23) and east sides of the enclosure. It is satisfying that part of the eastern shelter belt (Plate 24) was planted on the site of the spinney which existed in Braziers Close in the 1830s. The hedge of the northern shelter belt of the Mere is more or less on the line of the hedge which once separated Braziers Close from Boxworth End Close (see page 24). I needed shelter quickly so I planted fast-growing trees on the outer edges: 15 hybrid black-poplars alternating with 15 red alders. Inside them I planted trees which would grow more slowly – initially oaks, ashes and willows. Later I added a Lombardy poplar, ten hornbeams, two sweet chestnuts, an elm, three birches, a small-leaved lime and a black-poplar. The last was a kind present from Jim Bingley. He had been given it by Edgar Milne-Redhead, the expert on this rare tree, so I could be certain it really was a true native black-poplar. Initially the hedges consisted of a hundred hawthorns and a hundred blackthorns, the blackthorns being confined to the eastern and southern hedges. Finally, to give shelter before the shelter belt could provide it, I planted quick-growing elders along the top of the northern bank and, with an eye to the distant future, yew seedlings which had sown themselves in our garden.

The poplars did well: the Lombardy and 12 of the 15 black-poplar hybrids survived and grew fast. All the red alders died. I had hoped that this species could thrive on land which was too dry for the native alder, but I learnt the hard way that it could not. There were some casualties among the other trees. Two or three oaks rocked in the wind and had to be staked with willow stakes. One of them died but its willow stake took root and flourished! The one elm – a sucker from the Main Plantation – grew well initially, but when the bark thickened the beetles found it and it developed Dutch Elm Disease. When the trees were young I watered them when it was very dry, obtaining water from the Mere nearby. Otherwise I had little management to do. A list of the tree species surviving in the Mere Enclosure in 1997 is shown in Table 2.

The hawthorns in the hedge did very well, apart from two which developed fire-blight and had to be replaced. Blackthorn was much harder to get established, however many did survive and the southern hedge is now a wonderful sight in early spring when it is white with blossom. Where necessary I planted hawthorns in the gaps, also two buckthorns, one of which survived and I hope will provide food for brimstone butterfly caterpillars.

So far the hedges have required little management, although fairly soon they will need topping to encourage a thick base which is important for nesting birds. The blackthorns have sent many suckers out into the grassland. Those on the field side get grazed down by cattle into compact little bushes. I have to cut those on the north side to prevent them invading the Mere meadow.

The Mere filled up remarkably quickly and by the spring of 1984 it was completely full (Plate 25). During that spring and early summer I began to introduce water plants. It was difficult to predict which would do well so I tried out a wide range of species. Many of them grew in nearby ditches and gravel pits but these were very different habitats from the Mere. I was helped by kind Monks Wood colleagues: Colin and Joan Welch gave me many plants from their fine pond in Northamptonshire. The species introduced into the Mere are listed in Table 5 together with those which colonized it naturally or, just conceivably, were accidentally introduced with other species. Most of those which were introduced remained, and in some cases spread. The water was slightly acid which may explain why water-soldier and frog-bit eventually disappeared. On the other hand bogbean may have died out because the water was not acid enough. Some species do better in some years than in others. Floating pondweed fluctuates noticeably.

The species which have been most successful are the ones that had apparently got there by themselves: reed, bulrush and the alien New Zealand pygmyweed, *Crassula helmsii*, which had been introduced to Britain. Branched bur-reed, yellow flag (Plates 26 and 27) and greater spearwort, which I did introduce, have also spread well.

Ponds, whether natural or man-made, are usually very transient habitats. They quickly fill with debris, become marshes and, when these are dry enough, they become invaded by willows or alders and eventually become woods. Each stage from open water to damp woodland supports interesting plants and animals. If you are lucky enough to own or manage a large amount of land you can develop a programme of continually making new ponds so that the flora and fauna of all stages will be catered for, including those which interest you most. But for most people this is an impossible dream: you have only got one pond. As a result, you have to manage it if you want it to mark time at a particular stage of development. For most people, and for most dragonflies, that is the stage when there is plenty of open water and it is surrounded by marsh plants (see Plate 27). The Mere was at this stage for about ten years, but I knew that it would not remain so for ever. After about four years the Mere had been invaded by bulrushes and reeds. In 1988 there had been only 13 bulrushes and three reed plants. However by 1993 bulrushes occupied most of

Table 5     Success and failure of aquatic plants introduced into the Mere, 1984–2000

| Species | Subsequent development |
| --- | --- |
| **Submerged and Floating** | |
| † White Water-lily (*Nymphaea alba*) | Original plants have survived |
| † Pond Water-crowfoot (*Ranunculus peltatus*) | Survived where introduced |
| *† New Zealand Pigmyweed (*Crassula helmsii*) | Increased rapidly until drought in 1997. Still abundant |
| Frogbit (*Hydrocharis morsus-ranae*) | Disappeared within one or two years |
| Water-soldier (*Stratiotes aloides*) | Survived only about four years |
| Canadian Waterweed (*Elodea canadensis*) | Survived a few years only |
| † Broad-leaved Pondweed (*Potamogeton natans*) | Fluctuates greatly from year to year |
| Curled Pondweed (*Potamogeton crispus*) | Survived about two years |
| **Emergent** | |
| Bogbean (*Menyanthes trifoliata*) | Survived several years |
| † Mare's-tail (*Hippuris vulgaris*) | Survived in small numbers and has spread |
| † Flowering-rush (*Butomus umbellatus*) | Survived and spread |
| *† Common Reed (*Phragmites australis*) | Spread rapidly. Has to be controlled |
| *† Bulrush (*Typha latifolia*) | Spread rapidly. Has to be controlled |
| **Edge** | |
| Marsh-marigold (*Caltha palustris*) | Survived only a few years |
| † Greater Spearwort (*Ranunculus lingua*) | Survived and is spreading |
| † Creeping-Jenny (*Lysimachia nummularia*) | Has spread, especially since drought years |
| † Water-plantain (*Alisma plantago-aquatica*) | Increased rapidly until drought years |
| † Soft-rush (*Juncus effusus*) | Survived |
| † Sedges (2 species) (*Carex* spp.) | Spread especially since drought years |
| † Branched Bur-reed (*Sparganium erectum*) | Spread rapidly until drought year of 1997 |
| † Yellow Iris (*Iris pseudacorus*) | Survived and spread |

Notes: (i)   All species were intentionally introduced except those marked * which apparently colonized the Mere unaided.
   (ii)  † Species present in 2000.
   (iii) † Stonewort (*Chara*), a lower plant, has also colonized the pond.

the narrow eastern end of the Mere and reeds and bulrushes extended out to the island on its northern and western shores. That year Caroline gave me a little boat as a 70th birthday present and I began to control the reeds from the boat by cutting them below the surface of the water (Plates 28 and 29). The exceptional drought of 1996 lowered the water-level to such an extent that reeds were able to spread into the deeper water on the south side of the island: they threatened to take over the whole Mere. There are few more beautiful things to see and hear than a reed bed moving in the wind, but reeds shade out submerged plants and restrict the amount of sunlight which reaches the water so they are bad for dragonflies in the cool climate of Britain. As the main purpose of the Mere is to provide a good habitat for dragonflies, I have to continue to control the reeds. To do this by hand is a time-consuming business but it has to be done.

The grassland round the Mere became the happy hunting ground for dragon-flies, as predicted, but I also wanted it to be a place for butterflies (Plate 30). As for dragonflies, one has to cater for their larvae as well as for the adults. Most of the caterpillars of the butterflies which occur in the Swavesey area feed on grasses, crucifers or nettles. All these plants were present in the Mere Meadow when it was enclosed, or soon appeared when it was no longer grazed. However, initially it did lack nectar-bearing plants required by adult butterflies. Soon an increase in creeping and spear thistles improved the situation, but I decided to improve it further. I planted brambles in sunny places on both the northern and southern banks. They did well, not only providing many flowers for butterflies but also blackberries for birds (and us). I also planted teasels for red admirals and other butterflies and also for goldfinches, which feed on their seeds. The holly blue was catered for by holly and ivy in our wood and garden, but I hoped we could also get the common blue to breed if we could provide bird's-foot-trefoil for its caterpillars. Sadly this plant does not like heavy clay and I was unable to get the common blue established as a breeding species although it frequently visits the Mere Enclosure. However, the brown argus, which can feed on plants of the family Geraniaceae, recently established itself for two years.

By 1983 the flora of the field had been made more agriculturally productive but had become poorer in species. Therefore I decided to enrich the flora of the Mere Enclosure. I thought coltsfoot would do well on the clay banks, but it did not. I have already mentioned (see page 41) that cowslips planted in the Mere Meadow eventually died out, although they had occurred originally in the field when it was lightly grazed. Yarrow, mugwort, buttercups, celandines, stinking iris, great burnet, primroses and violets all became established where I had planted them. Oxeye daisy and meadow crane's-bill not only became established on the southern bank but seeded themselves and spread. Several species came in without my help – cow parsley, hogweed, hemlock, bristly oxtongue, wood woundwort, and the rather rare rough hawk's-beard. This species stayed only for a year or two before disappearing. Today the grassland of the Mere Meadow and the northern and southern banks is considerably richer in species then when it was first enclosed in 1984 and, as we shall see, supports many butterflies.

Grassland on heavy clay is remarkably persistent even when it is not grazed. However, as the hedges and trees grow up, thistles, oxtongues and umbellifers increase and so reduce the amount of ground covered by grasses. Worse still, suckers from the blackthorns in the southern hedge threaten to turn the Mere Meadow into scrub if they are not controlled. Therefore to maintain grass-feeding butterflies I must pull up an increasing number of thistles and cut the blackthorn suckers each year. I also mow paths inside and outside the banks and occasionally I mow the rest of the Mere Meadow.

Most of the plants which occur in the Swavesey Wood or in the Mere and its surroundings were planted by me; rather few plants have colonized the area on their own. However, unlike plants, many animals did, and in the next chapter I shall describe them and how they colonized the different habitats I had provided as they developed.

# CHAPTER EIGHT
# COLONIZATION BY ANIMALS

A. *Introduction*

This chapter is about the colonization of my nature reserve – its wood, mere and grassland – by different sorts of animals. What is described is based on notes which were made at the time throughout the forty-year period. The amount of time I could spend making observations varied greatly from year to year, therefore the records are uneven. Nevertheless the general picture of what occurred is quite clear. The new habitats, once created, developed naturally but they were modified by any management which I undertook, for example by mowing grass and cutting reeds. Even in the wood I weeded cow parsley from the woodland flower plots and cut off dead branches from trees when they looked hazardous. Also there was interaction between the different habitats: birds habitually used the Mere as a drinking place and they fed in its surroundings. The whole reserve had interactions with the garden and field which pre-dated the Wood and Mere. The field to the north of our field which had been allocated to Mr Wragg at the Enclosures became a mobile home site.

The reserve was affected by things happening outside it. For example, agricultural practices in East Anglia changed over the years. More grassland came under crops. The population of Swavesey doubled. This was not necessarily bad for wildlife as more houses meant more gardens: nowadays gardens often support more species than arable land or improved grassland. Events much further afield must also have impinged. For example, as we shall see in the case of the whitethroat, migrant species could be affected by changes in land use or climate in their winter quarters in Africa.

In the next four sections I describe my observations on the groups of animals which I have studied. In the last section I comment on the far more numerous species in the reserve which I have not studied.

## B. *Birds*

Of all the animals one sees in Britain birds
are the most popular and the easiest to
identify, so I shall start with them.

The stages of development of our small nature reserve (see Figure 4) that
determined which species colonized it were as follows:

1. 1961–64    Part of a grass field
2. 1964–71    A small wood of about a third of an acre (the Buckingway
Plantation linked to the Old Plantation by the Boundary Shelter
Belt)
3. 1971–83    A larger wood of about one acre (the Buckingway Plantation,
the Boundary Shelter Belt, the Old Plantation and the Main
Plantation)
4. 1983 –     The larger Wood (see 3 above) joined to the Mere with its
Shelter Belt and Meadow, the whole covering about 1.6 acres

The nearest source for colonists of the Swavesey Wood was our garden which
adjoined it. When we arrived at The Farm House in 1960 the garden supported
13 bird species: wood pigeon, wren, dunnock, robin, blackbird, song thrush,
blue tit, great tit, starling, house sparrow, chaffinch and greenfinch, which
were all residents, and the swallow, which was a summer migrant. These
species, except for the swallow, have continued to be present in the garden
practically every year for 40 years. In 1970 they were joined by the collared
dove, making a total of 14 regularly breeding species. Spotted flycatcher,
goldfinch and jackdaw breed occasionally and in 1970 tree sparrows nested in
a bird-box in the orchard.

The garden is frequently visited by bullfinch and carrion crow, previously by
yellowhammer, and since 1989 by sparrowhawks. Mistle thrushes sometimes
come to feed on yew berries. Much rarer visitors have been green woodpecker,
great spotted woodpecker, blackcap, willow warbler, goldcrest, pied flycatcher
and treecreeper. Several other species were seen flying high overhead: cuckoo,
heron, mute swan, mallard, golden plover, lapwing, gulls, rooks and starlings
and once a hobby. By the end of 2001 we had seen a total of 82 species on or
over our land (see Appendix 2).

All the breeding species in the garden, except for house sparrows and
swallows, are essentially woodland species and it was not surprising that none
of them colonized the plantations at an early stage. In fact the early stages of all
the plantations were singularly devoid of birds. I thought that both whitethroat
and yellowhammer might colonize the Main Plantation at least during its scrub
stage. I suspect the whitethroat would have done so had not the time when the
Main Plantation was most suitable for it coincided with the drought in the

Sahel zone of Africa, where the species spends its winter. The exceptional drought in Africa caused the whitethroat population in Britain to be reduced to less than half of its normal numbers. As a result the remaining whitethroats appeared to cling to their established sites and tended not to colonize new ones. In 1968, when whitethroats were still abundant, we had a pair in the eastern hedge of our field. The following year it did not return, and it was not until 1992 that we had a whitethroat again. It colonized the scrub by the Mere. Although I often saw yellowhammers in our garden in the springs of the 1970s and early '80s, I saw it only one year in the Main Plantation (1976) and, so far as I know, it did not breed there. Clearly the habitat was less suitable for the species than it appeared to be.

The true woodland birds began to colonize the Main Plantation when the trees were about 13 years old and 15–18 feet high. Blackcap, willow warbler and tawny owl arrived in 1985 but turtle dove, robin, blackbird, song thrush, lesser whitethroat, spotted flycatcher, long-tailed tit and chaffinch not until 1989, when the trees had been there 17 years. Since then the only new breeding species in the wood have been the magpie which first nested in 1991 and the rook. Rooks were seen prospecting the Old Plantation in October 1995, and in 1996 they set up two rookeries, each of six nests, one in the Buckingway Plantation (Plate 31) and one in the Old Plantation (Plate 32), the two parts of the Swavesey Wood which contain the oldest trees – those planted 35 years ago. With the coming of the rooks I felt that the wood really had come of age. Although rooks scatter twigs all over the lawn when they build their nests we are glad to have them as neighbours. We can watch the boisterous inhabitants of both rookeries from the windows of our bedroom without even getting out of bed.

The Swavesey Wood now has 12 bird species which breed regularly or frequently. Eight of them (wren, dunnock, robin, blackbird, song thrush, chaffinch, wood pigeon and collared dove) are resident species which also breed in our garden. In addition four species which have never bred in our garden have colonized the wood. Of these, tawny owl and rook are also resident species but blackcap and willow warbler are summer migrants. In addition two other residents, magpie and long-tailed tit, sometimes breed, and the turtle dove, a summer migrant, has bred at least once.

Sparrowhawks have hunted in the wood since 1989. They frequently kill collared doves. Jays visited it in the period 1990–93. Since 1991 pheasants have roosted in it. When there are hawthorn berries on the boundary hedge of the Main Plantation fieldfares and redwings come and feed on them. These beautiful migrant thrushes from north-eastern Europe also feed on invertebrates in our field. One day in November 1989 I had the pleasure of putting up a woodcock from the one damp spot in the Wood – the only place there where an alder has survived (see Plate 2).

The trees are still too young to attract breeding treecreepers, nuthatches or woodpeckers, but I have seen a great spotted woodpecker twice in the wood and several times in the garden, a treecreeper in the garden, and green woodpeckers increasingly in the garden and by the Mere. Perhaps eventually one or more of these tree-climbing species will colonize the Wood.

In recent years much research has been done on the birds of islands, and it has been found that there is a fairly constant relationship between the number of species breeding and the size of the island. Small woods are rather like islands in a sea of agricultural land. In the 1970s Max Hooper and I studied the birds of different sized woods and found that a similar relationship occurred to that found in the geographical islands. Very roughly this relationship can be expressed thus: islands ten times the area of other islands have twice their number of species. I was naturally interested to see whether increasing the area of my little wood would bear out this proposition. Originally the Buckingway Plantation together with the Boundary Shelter Belt and the Old Plantation made up a small wood of about a third of an acre. It contained four bird species – blackbird, song thrush, dunnock and robin. To double this number of species the Swavesey Wood would have to be ten times the area, that is about three acres. In fact the total Swavesey Wood was only one acre. I had done little more than treble the size of the original wood by adding the Main Plantation to the older plantations, yet the whole wood now has 12 species; that is three times the original number. How can one account for this apparent anomaly? I suspect that the reason was this: the Swavesey Wood, unlike the woods studied by Max Hooper and myself, is not completely isolated; it is adjoined by trees and bushes in neighbouring gardens including our own (see Figure 3, p. 18). Therefore the Swavesey Wood is not a true 'island' as it would be if it were entirely surrounded by arable or improved pasture. Consequently the well-established relationship between island size and its bird fauna could not be used to predict the number of its bird species.

Birds are territorial and so the number of pairs in a small wood is consequently low. There have never been more than two pairs of blackcap, willow warbler or chaffinch in the Swavesey Wood. Even blackbirds rarely exceed four pairs. Only the colonial rook gets into double figures: there were 12 nests altogether in the two rookeries in 1996. By 2001 there were 25.

The size of water-bodies also determines the number of species found on them. Great crested grebes, little grebes, tufted ducks and coots all occur on the Swavesey gravel pits but they all depend on food obtained from the water whether it is waterweed, invertebrates or fish. A pond like the Swavesey Mere which is only about 300 square yards in size could not sustain any of these species. The smallest pond which I have ever known to support a breeding coot was 1250 square yards, and the smallest for a great crested grebe 17,940 square yards. Moorhens and mallard, while preferring to nest by water, are not entirely dependent upon it for food and so can breed in or by much smaller ponds. I was fairly confident that these two species would find the Mere and breed there. They did. A pair of mallard appeared in 1985, 18 months after the Mere had been dug. They first bred in 1989. Then and subsequently they have always bred on the island. Sometimes their eggs are taken by a magpie or a crow, but sometimes they escape and I see ducklings swimming on the Mere or hiding in its water plants. Usually they disappear fairly quickly: perhaps their mother takes them to a safer place or perhaps they are eaten by a fox. Moorhens appeared in 1986, and like the mallard, first bred in 1989. Sometimes their eggs too get taken, but at other times they hatch and the chicks remain on the Mere until they are nearly full grown.

The Mere is too small to support more than two breeding water birds but it has attracted some very welcome visitors. The appearance of sedge warbler, reed warbler and reed bunting in recent years has made it clear that I must control reeds more effectively if I am to keep the dragonflies! Yellow wagtails have been visiting the Mere since 1985 and their visits are always welcome even if these beautiful birds do come to eat emerging damselflies. In 1990 the Mere was paid a short visit by a kingfisher and in 1997 by a teal. Snipe began to visit the Mere once it was filled with water. Since then they have appeared most years at various times from August to May. Herons first visited the Mere in 1989. As there were no fish until 1997 they presumably come, generally in autumn, to catch newts and invertebrates.

Figure 8: Spoonbill on the north shore of the Mere, 12th May 1990

In 1990 the Mere received a most distinguished visitor. We had been to Cambridge to see our nephew Tom Oliver get his degree. After the ceremony he and his family came home to tea. Tom is a keen naturalist with all the right priorities. I was therefore not surprised when he said, 'First we must have a look at the Mere'. So he and I, not expecting more than a possible mallard or moorhen, hurried through the Wood and out to the Mere. There, standing like a white statue beside it, was a spoonbill! (Figure 8) We crept away to tell the others and fetch a camera. The spoonbill obligingly remained where it was, and was cautiously admired by the whole party. It stayed for about an hour and then flew off. Next day we saw it flying over the large gravel pit which straddles the Swavesey parish boundary. I had only once seen a spoonbill before in Cambridgeshire. Oddly enough it was only about two weeks previously. Unlike the Swavesey one, it was a young bird with black tips to its wing feathers. For quite a time afterwards we dined out on our spoonbill story. I got the impression that quite a lot of people thought we had invented it, but we had not!

## C. *Mammals, Reptiles, Amphibians and Fish*

Mammals are much harder to observe than birds, and I have to admit that I owe much of what follows to Tinkle, Caroline's cat who, while collecting mammals for her own sake, graciously allowed me to identify them. At one time I thought I would publish a short paper about our Swavesey mammals; if I had done so I would have felt obliged to give Tinkle the status of co-author. Alas, I did not write the paper, Tinkle never became an author and was eventually eaten by a fox.

Thanks to Tinkle's hunting skills and my observations, I know that our house and garden supported at least eight species of mammal: hedgehog, common shrew, brown long-eared bat, pipistrelle bat, bank vole, wood mouse, brown rat and house mouse. Tinkle also caught field voles and a harvest mouse, but she had probably caught these two species in the field, not in our garden. I have only once seen a live harvest mouse in Cambridgeshire. It was climbing in the hedge which runs into the eastern hedge of our field.

I expect foxes have quite often visited our garden but I never saw one until I found a dead one in the Main Plantation in 1976. On 28th September 1987 I saw a young fox, about two-thirds grown, by the Mere. For a time a fox lived under a pile of dead wood between the Buckingway and Old Plantations. In the '90s I saw foxes by the Mere quite frequently. They had an earth under the large bramble bush on the southern bank (see Plate 30) and in 1997 reared three cubs. They appeared to collect a miscellany of things to play with including dead birds, bones and discarded children's toys. The only other wild predator I have seen is the stoat. The first one I saw was in the Main Plantation in 1976. In 1998 there were clear signs that a badger was digging holes by the Mere, but so far I have not seen him or her.

Surprisingly I have never seen signs of moles in our garden, nor are they common in the field. However, from 1989 onwards, the Main Plantation with its leaf litter appeared to provide a habitat to their liking. I saw the first rabbit in the Main Plantation that same year. Since then I have seen them both there and by the Mere where the bramble bush conceals a burrow. Grey squirrels arrived in 1987, since when they have been present most years in the garden and Wood. Their drey has always been in the Main Plantation part of the Wood. Before the squirrels arrived we used to get good crops of nuts from the hazels I had planted; now we get none. Hares need wide open spaces and are a characteristic species of the large arable fields of Cambridgeshire. Their populations fluctuate greatly and sometimes hares are extremely rare. In the early days I used to see them in our field from time to time, and in 1986 I nearly trod on a leveret in the middle of it. Recently I have seen none.

Undoubtedly the best place for mammals in our reserve is the Mere and its surrounding grassland. I have seen the following 11 species there, dead or alive: pygmy, common and water shrew, small bats (probably pipistrelle), rabbit, bank, water and field (or short-tailed) vole, brown rat, fox and stoat. The water voles bred on the island in 1986. Their four babies were delightful to watch as they bumbled about in the water-weed. I saw water voles again in 1987 but have seen none since. Brown rats colonized the island at that time and were probably the cause of the disappearance of the water voles. The stoat which appeared in 1992 was a young animal and did not stay. Doubtless foxes prey on the other mammals of the Mere enclosure, but I suspect that the main predators there are tame cats from neighbouring houses.

To conclude, I have recorded 19 species of mammals in our garden and small reserve – that is 40 per cent of the native fauna – as well as the two well-established introduced species, the rabbit and the grey squirrel. Sadly the two aquatic mammals, the water vole and water shrew, seem unable to maintain themselves by the Mere. An inexplicable loss in the last two years has been the hedgehog. The habitat appears to be improving for them all the time yet they seem to have deserted us at least temporarily.

Lizards and snakes are very much localized in Cambridgeshire and it is not surprising that it was a long time before we recorded a reptile. On 15th July 1998 a friend and I saw a two-foot grass snake slide from the north shore of the Mere into the water and swim to the edge of the island. It lay motionless in the water for a while before disappearing into the reeds. I saw it again a month later. It, or another grass snake of similar size, appeared on 15th June the following year and was last seen on 16th July 1999. I saw it or another on 19th May 2000. On 7th June that year I nearly trod on a much smaller grass snake which was lying on the grass by the water's edge. It was about one foot long and very tame. Shortly afterwards I saw another small snake at the other end of the Mere. One of them was present next day, but I did not see a small snake after that date. The large snake remained: I last saw it on 21st September 2000. It looks as if the grass snake breeds nearby, possibly in one of the piles of cut reeds, etc. which I leave for the purpose near the Mere.

We have done better with Amphibia, having recorded four species. I have mentioned above that a great crested newt discovered an old tin bath which we had put in our garden near the back door and kept full of water. Later when we made the butyl-lined pond in the enclosure between the Buckingway and Old Plantations it quickly became colonized by smooth newts. I often unearthed them when digging in the garden. I tried to introduce frogs into the pond, but with total lack of success. I soon discovered the reason. One night I left a bowl full of tadpoles beside the pond intending to put it into the pond next day. When I visited it next day there were no tadpoles, but no fewer than 14 well-fed smooth newts were attempting to get out of the bowl! When the pond dried out the newts disappeared, but in 1985 they found a new habitat in the Mere. Since then the Mere has usually supported a large population of smooth newts. They breed successfully every year. They are best seen on warm spring days when they keep coming to the surface to breathe. Despite my failure at the pond, I also introduced frog spawn into the Mere. Subsequently I have rarely seen frog spawn, but some tadpoles may have survived because I see frogs in or by the Mere most years and in the late summer and autumn I quite often see them in the garden. In 1994 and 1995 I met a toad there too.

April 29th 1999 was a red-letter day because, while watching smooth newts rise to the surface for air and pursuing their prey, I saw two much larger newts on the south side of the Mere. They were female great crested newts. I also saw what I was pretty sure was a male newt of the same species at the west end of the Mere. Four days later I watched a female great crested newt trying with only partial success to eat a great pond snail (*Lymnaea stagnalis*) which was floating on the water surface. I got another very good view when a female surfaced to get air on 13th July. On 19th August that year I was clearing some old bricks from underneath a box hedge in our garden and found three great crested newts hiding in holes in the bricks and underneath them. This strongly suggests that our great crested newts travel at least 160 yards between their breeding place and their out-of-water habitat. I also found a small one under a tussock of grass at the foot of a plum tree, which suggests that they have already bred successfully in the Mere. And so it was that in 1999 we properly welcomed our first Red Data Book species to our reserve – a species which is protected both by British and European legislation. Sadly I saw no great crested newts in 2000 and smooth newts were much rarer that year than usual. In the absence of frogs, the grass snakes are probably feeding extensively on newts. As welcome as the grass snakes are at the Mere, they may be threatening our newts. It will be interesting to see if we can retain all three species.

Newts, toads, grass snakes and probably some of the frogs must have found their way to the Mere from farm and garden ponds and ditches in the neighbourhood. I dug up a great crested newt in our rose bed in 1997, two years before I saw one on the Mere. It may have been one of the pioneer colonists on its way to its new habitat.

For 13 years no fish were observed in the Mere but, on 23rd September 1997, my son, Peter, and grandson, Paul, noticed an unusual movement among the water-weed at the north-east part of the Mere. A closer look revealed two small fish, probably sticklebacks – our first fish. None have been seen since, which is probably good news for the dragonflies.

## D. *Butterflies*

Butterflies are the first love of most entomologists. Although many of us get intrigued with other groups, butterflies remain a delight for us all. So, if we can, we want to have as many species on our doorstep as possible. To achieve this we have to cater both for caterpillars and adult insects.
The caterpillars of butterflies are much fussier about what they eat than most birds, and even the adults have strong preferences. To encourage butterflies one must provide the right plants in the right places: small tortoiseshell caterpillars will not eat any old nettle, nor will orange-tip caterpillars eat any old crucifers.

Like our bird fauna, our butterfly fauna consists of both residents and migrants from abroad. Unlike birds, some of the latter, such as red admirals and painted ladies, come here only to die, because they only very rarely survive our winters. All our other species, whether resident or migrant, survive by hibernating: as adults, such as small tortoiseshells and brimstones; as pupae, such as the holly blue; or as larvae, such as the meadow brown. As well as food and shelter we must also provide places in which butterflies can hibernate.

The changes of the butterfly populations which I have observed in our reserve mainly reflect the changes of habitat which I have produced by planting and management. For butterflies the most important have been the changing of grazed to ungrazed grassland and the changing of grassland to woodland. The digging of the Mere has had no effect since no British butterfly is aquatic.

However, a few moths – the china-marks – have aquatic larvae which feed on water plants. One species even has a form of the adult female which is also aquatic. I have occasionally seen china-mark moths on the Mere.

To study changes in butterfly populations one needs to be able to assess their numbers. By the time we moved to Swavesey in 1960 several studies had been made on populations of butterflies. Their numbers had been estimated by marking individuals and recording how often the marked insects had been recaptured. This is a laborious process and many years ago I wondered if there was an easier way to assess trends in butterfly populations. For some time I had used transects to study adult dragonflies: I counted the numbers of males which I saw while walking along the edges of streams and ponds. I found that so long as I restricted my observations to times when conditions were optimal for dragonflies – that is about mid-day on fine days when there is little wind – the results were comparable, and so I could assess changes in dragonfly populations during the season, and I could compare season with season. In 1963 the Nature Conservancy's Monks Wood Experimental Station had been completed and I had gone to work there. I noticed that the southern edge of the wood which adjoins the laboratories was particularly good for butterflies. In 1964 I decided to see whether my method of counting dragonflies could be applied to butterflies. We had breaks at mid-day and if one did not spend too long over lunch there was often time to walk along a five-hundred metres stretch of the south edge of Monks Wood counting the butterflies as I walked. I did this for ten seasons and came to the decision that while the method 'is too insensitive for accurate population work it has considerable potential for assessing the value of different habitats for butterflies and for detecting order of magnitude population changes'. Meanwhile my colleague, Ernie Pollard, who was studying the effects of pesticides on insects in farmland, became interested in what I was doing, made his own studies and made refinements to my methodology. In 1975 we published the results of our studies in the *Entomologist's Gazette*. From these small beginnings, Ernie Pollard developed the Butterfly Monitoring Scheme which has proved to be one of the most enduring and successful indicators of changes in the status of wild animals in Britain.

When we arrived in Swavesey in 1960 I had not studied the efficacy of butterfly walks and only undertook them tentatively on our land. However, by 1970 I had sufficient confidence in the reliability of butterfly transects to try them out in the old enclosure during that and the following year (see page 31). The results were promising and I was determined to do more butterfly transects when I had retired and the reserve had been enlarged. In the event I was able to start the transects in 1982, two seasons before the Mere was constructed. I have continued these transects in the garden, in the Swavesey Wood and in the Mere Enclosure ever since. When making butterfly transects correct identification is crucial. Most British butterflies can be identified on the wing once one is familiar with them. Others can be identified for certain only if you can see the undersides of their wings, so these species must be observed while at rest. Green-veined and small whites and female orange-tips come into this category. Care must also be taken to distinguish between large whites and female brimstones which can look very similar in flight. Finally there is a pair of species which are so alike that they have to be scrutinized, ideally in the hand, if they are to be distinguished from each other. They are the small and Essex skippers (Plate 33). Considerable gymnastics have to be performed if one is to determine whether the underside of the tip of the antenna is black – the hallmark of the Essex skipper; it is often easier to catch the insect. Whenever possible I identified the species of every butterfly seen, but sometimes when pushed for time I merely recorded 'whites', i.e. green-veined or small whites or female orange-tip, and 'small/Essex skippers.'

So what happened over the years in the Swavesey Wood and the Mere Enclosure? Table 6 shows how the butterfly fauna of the Swavesey Wood changed from 1982 to 2000. In 1982 woodland was confined to the Buckingway and Old Plantations. The Main Plantation was still a mixture of small trees and grassland. By 2000 the whole area consisted of woodland except for a small area of open ground by the site of the old pond. It will be seen that the gatekeeper (Figure 9), meadow brown and ringlet, whose caterpillars feed on grasses, all declined. Small whites, small tortoiseshells and peacocks also declined, probably because the flowers on which the adults feed became fewer as the wood developed. The only new species to colonize the wood was the speckled wood (Plate 34), a true woodland species, which appeared in 1992. Of the remaining 13 species, the green-veined white, orange-tip and holly blue continued to appear in small numbers most years throughout the period, others more rarely. Yet others such as the large skipper, were probably only occasional transient visitors from the garden or the field. In 2000 only three species were recorded.

Table 7 shows how the butterfly fauna of the Mere Enclosure developed from the time it was enclosed in 1984 until 2000. In the first year it was colonized by 17 species. Of these, some 11 species increased in numbers in later years; the remaining six remained in low numbers. The development of the populations of the four grass-feeding, brown species of butterfly in the Mere Enclosure is illustrated in Figure 10, which can be compared with what happened to these species in the Main Plantation (Figure 9). The provision of rough grassland appears to have led to increases in the populations of gatekeeper, meadow brown and ringlet. Later, when the area of sun-exposed grassland became reduced by the growth of hedges, all these species declined. The speckled wood colonized the shelter belt ten years after it had been planted. After 1984 five more species were observed in the Mere Enclosure, but usually

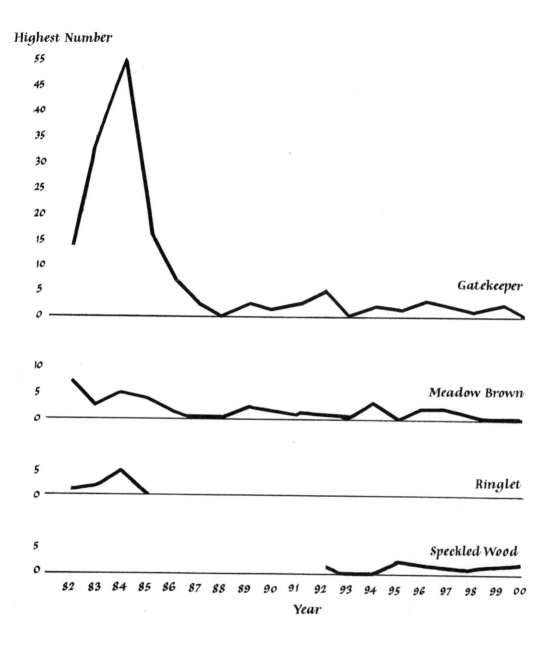

**Highest Number**

Gatekeeper

Meadow Brown

Ringlet

Speckled Wood

Figure 9: Changes in the populations of grass-feeding butterflies in the Swavesey Wood as it developed from grassland to closed canopy woodland, 1982–2000

Figures are based on butterfly transects (see p. 63). Transect length in the Main and the Old Plantation and Old Enclosure c.175m.

Table 6   Changes in butterfly populations in the Swavesey wood, 1982–2000

| Species | 1982 | 1983 | 1984 | 1985 | 1986 | 1987 | 1988 | 1989 | 1990 | 1991 | 1992 | 1993 | 1994 | 1995 | 1996 | 1997 | 1998 | 1999 | 2000 |
|---|---|---|---|---|---|---|---|---|---|---|---|---|---|---|---|---|---|---|---|
| | Largest number of insects observed on one transect in the year indicated | | | | | | | | | | | | | | | | | | |
| Small Skipper | 1 | | | | | | | | | | | | | | | | | | 2 |
| Essex Skipper | | | 1 | | | | | | | | | | | | | | | | 1 |
| Large Skipper | | 1 | 1 | | | | | | | 1 | | 1 | | | 1 | 1 | | | |
| Brimstone | | 1 | | 2 | 1 | 1 | | | | | 3 | 1 | | 1 | 1 | 1 | 1 | 1 | |
| Large White | 3 | 4 | 1 | 1 | 1 | 1 | 2 | 1 | 1 | 1 | 3 | 1 | 1 | 1 | 1 | 1 | 1 | | |
| Green-veined & Small Whites* | **16** | 4 | 4 | 9 | 5 | 3 | 2 | 1 | 1 | 2 | 4 | 5 | 3 | 6 | 7 | 8 | 4 | 4 | 1 |
| Orange-tip | 3 | 3 | 2 | 2 | 1 | 1 | 1 | 1 | 2 | 1 | 2 | 2 | 1 | 1 | 2 | 1 | 1 | 1 | |
| Small Copper | 1 | | | | | | | | | | | | | | | | | | |
| Holly Blue | | 1 | 2 | 2 | 1 | | | | 4 | 3 | 4 | 1 | 1 | | 3 | 4 | 1 | 1 | |
| Red Admiral | 6 | 1 | 1 | 2 | | 1 | | 1 | 1 | 1 | 1 | 1 | | 2 | 1 | | | | |
| Painted Lady | 1 | | | | | | | | | | | | | | 1 | | | | |
| Small Tortoiseshell | 9 | 1 | 3 | 1 | 1 | | | | | 1 | 1 | | 1 | 1 | 4 | 1 | 1 | | |
| Peacock | 9 | 7 | 2 | 2 | | | | | | | 1 | 1 | 1 | 1 | 1 | 1 | 1 | | |
| Comma | 1 | | 1 | | 1 | | | | | 1 | 1 | 1 | 1 | 2 | 1 | 1 | 1 | 2 | 2 |
| Speckled Wood | | | | | | | | | 1 | | 1 | | | | | | | | |
| Wall | | 2 | 1 | | | | | | | | | | | | | | | | |
| Gatekeeper | 13 | 36 | **54** | 16 | 7 | 2 | | 2 | 1 | 2 | 5 | | 2 | 1 | 3 | 2 | 1 | 2 | |
| Meadow Brown | 7 | 2 | 5 | 4 | 1 | | | 2 | 2 | 1 | 1 | | 3 | | 1 | 1 | 1 | | |
| Ringlet | 1 | 2 | **5** | | | | | | | | | | | | | | | | |

Notes: (i)   * Both species were observed in all but three years. The records may include a very few misidentified female Orange-tips (see p. 64)

(ii)   The largest number of Green-veined Whites counted was 7 in 1985.

(iii)   The largest number of Small Whites counted was 16 in 1982.

(iv)   Peak numbers of more numerous species are shown in bold.

(v)   Transect (c.175m) in Main and Old Plantations and Old Enclosure.

Table 7   Changes in butterfly populations by the Mere, 1984–2000

| Species | Largest number of insects observed on one transect in the year indicated | | | | | | | | | | | | | | | | |
|---|---|---|---|---|---|---|---|---|---|---|---|---|---|---|---|---|---|
| | 1984 | 1985 | 1986 | 1987 | 1988 | 1989 | 1990 | 1991 | 1992 | 1993 | 1994 | 1995 | 1996 | 1997 | 1998 | 1999 | 2000 |
| Large Skipper | 1 | 2 | 8 | **20** | 4 | 3 | 2 | 4 | 4 | 5 | 10 | 7 | 7 | 4 | 3 | 10 | 3 |
| Small & Essex Skippers† | 2 | 4 | 5 | 13 | | 1 | 1 | 9 | 14 | 1 | 7 | 7 | 16 | **24** | 6 | 5 | 1 |
| Brimstone | 1 | 1 | 1 | | | | | 2 | 1 | 1 | 1 | 1 | 2 | 1 | | 2 | |
| Large White | 5 | 4 | 4 | 2 | 10 | 1 | 2 | 2 | **14** | 3 | 2 | 1 | 1 | 2 | 2 | 2 | 2 |
| Green-veined & Small Whites* | 7 | 9 | **80** | 22 | 30 | 12 | 8 | 13 | 31 | 7 | 12 | 15 | 17 | 14 | 16 | 10 | 16 |
| Orange-tip | | | 2 | 1 | 2 | 3 | 2 | **6** | 4 | 2 | 2 | 1 | 2 | 2 | 3 | 3 | 2 |
| Small Copper | 1 | | | | | | 2 | 1 | 2 | | 1 | 1 | 1 | 1 | 2 | | |
| Brown Argus | | | | | | | | | | | | | | 1 | 1 | | |
| Common Blue | 1 | | 1 | | | | | | 2 | 1 | 1 | 2 | 4 | 2 | 1 | 1 | |
| Holly Blue | | 1 | 1 | 1 | | 1 | 1 | 3 | **5** | 1 | | 1 | 2 | **5** | 1 | 1 | 1 |
| Red Admiral | 2 | 4 | 1 | 1 | | 1 | 1 | 1 | 2 | | 1 | 1 | 2 | 2 | 2 | 2 | 1 |
| Painted Lady | | 1 | 1 | 2 | 2 | | | | 1 | | | | **18** | | | 3 | 1 |
| Small Tortoiseshell | 7 | 27 | 29 | **49** | 5 | 18 | 17 | 20 | 12 | 13 | 23 | 33 | 10 | 9 | 5 | 10 | 8 |
| Peacock | 6 | 5 | 1 | 7 | 3 | 1 | 5 | 5 | 26 | 11 | 8 | **50** | 17 | 8 | 21 | 30 | 22 |
| Comma | 1 | | 2 | 4 | 1 | | 2 | 1 | 1 | | | 1 | 2 | 4 | 1 | 3 | 4 |
| Speckled Wood | | | | | | 1 | | | | | 1 | 2 | 1 | 1 | 1 | 1 | 4 |
| Wall | 3 | | 1 | | | 1 | 3 | 1 | 3 | 1 | 1 | | | 1 | | | |
| Gatekeeper | 19 | 77 | 66 | 122 | 71 | 136 | 86 | **138** | 115 | 53 | 85 | 79 | 53 | 39 | 22 | 30 | 7 |
| Meadow Brown | 3 | 13 | 8 | 12 | 3 | **31** | 12 | 13 | 15 | 10 | 26 | 19 | 9 | 8 | 13 | 4 | 5 |
| Small Heath | 1 | 1 | 1 | 1 | | 1 | | 1 | | | | | | | | | |
| Ringlet | 1 | 1 | 1 | 2 | | 8 | 9 | **10** | 8 | 1 | 2 | 4 | 4 | 2 | 5 | 4 | |

Notes:   (i)   † Both species were identified in 1985, 1986, 1991, 1992, 1995, 1996, 1997, 1998 and 1999.
         (ii)   * Both species were identified every year. The records may include a very few misidentified Orange-tips (see p. 64).
         (iii)  The largest number of Green-veined Whites counted was 12 in 1988.
         (iv)   The largest number of Small Whites counted was 80 in 1986
         (v)    Peak numbers of more numerous species are shown in bold.
         (vi)   Transect length c.230m.

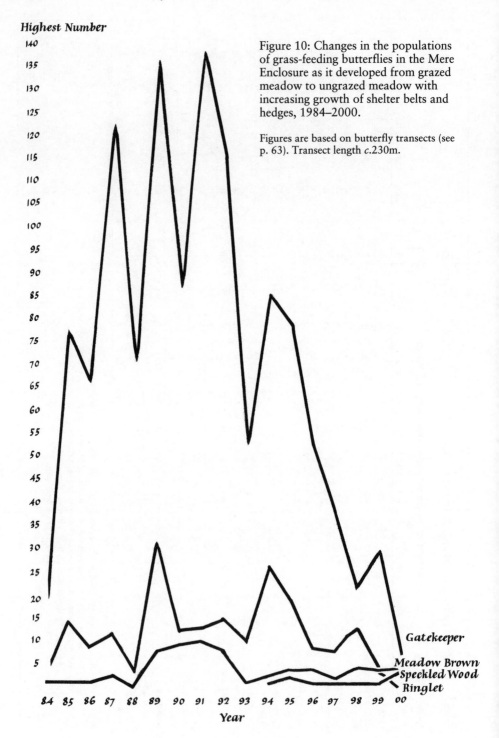

**Highest Number**

Figure 10: Changes in the populations of grass-feeding butterflies in the Mere Enclosure as it developed from grazed meadow to ungrazed meadow with increasing growth of shelter belts and hedges, 1984–2000.

Figures are based on butterfly transects (see p. 63). Transect length c.230m.

Gatekeeper

Meadow Brown
Speckled Wood
Ringlet

**Year**

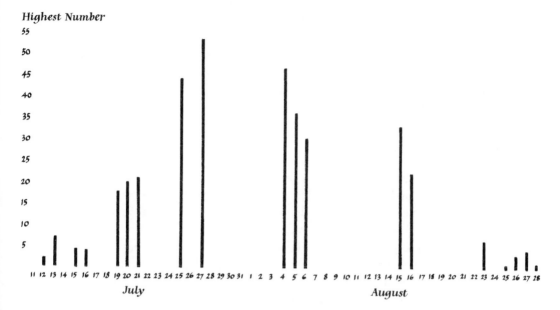

Figure 11: The rise and fall of the adult gatekeeper butterfly population in 1996
Figures are based on butterfly transects (see p. 63). Transect length c.230m.

only in small numbers. Table 7 also shows how different species peaked in different years. No fewer than seven species (large skipper, small and green-veined whites, small tortoiseshell, peacock, meadow brown and gatekeeper) were recorded every year. Fifteen other species have been observed in 12 to six seasons. Of these the speckled wood, first recorded in 1994, has been recorded in each of the last seven seasons.

Weather varied considerably throughout the period and this was reflected in the flight seasons of the butterflies. For example, in 1985 the flight season of the meadow brown was from 10th July to 30th August; in 1986, 30th June to 18th August; and in 1995, 23rd June to 13th August. It is interesting that the length of the season (50 or 51 days) was virtually the same in all three years. That of the more numerous gatekeeper was similar. Its gradual rise and decline in 1996 is shown in Figure 11. Some of the butterflies had more than one brood – notably the green-veined white, holly blue, wall and speckled wood.

Each year the same sequence occurs. In early spring small tortoiseshells, peacocks, commas and brimstones, which have hibernated as adults, appear on fine warm days. They are joined by green-veined whites, orange-tips and holly blues which have spent the winter as pupae. By June most of the spring butterflies have gone and there is quite a long period when there are usually very few butterflies about. The only species in Swavesey which flourishes at this time is the large skipper, although sometimes small and large whites are present too.

At this time of the year there are few grassland plants in flower that can be used by butterflies. Most are on the wing throughout July and the first half of August, when thistles and brambles and other flowers provide food for them. This is the height of our butterfly season. The large numbers round the Mere in midsummer bring back childhood memories of the time when most grasslands had not been improved. September produces late broods of walls and small coppers. Autumn is the time when the nymphaline butterflies, that is red admirals, peacocks, small tortoiseshells and commas, add lustre to the garden as they feed on Michaelmas daisies and rotten fruit.

While the general patterns of what has occurred are obvious enough, it is not easy to interpret the details. This is because several things were happening at the same time. Habitats were changing or being maintained by management and each species must have been affected by annual differences in weather, predation and parasitization. The migrant species were doubtless affected by weather and changes in land use in continental Europe and North Africa.

In the 1960s I recorded 16 species in our garden and field. In 1970 I added brimstone and white-letter hairstreak to the list; in 1971 holly blue; in 1982 comma, Essex skipper, small copper and painted lady; and in 1997 the brown argus. I would like to say that holly blues came because I planted hollies in 1976; brimstones because I planted buckthorn in 1985; and that the plants on the Mere embankments had brought in Essex skipper and small copper; but I cannot, because all these species appeared before I had done these things. What I hope I have done is to provide breeding places for species which were originally only visitors. It is almost certain that the woods have provided breeding places for the speckled wood which before was only a rare visitor. The loss of the white-letter hairstreak, our rarest butterfly, must be attributed to the loss of our elms on which the caterpillar fed. The rose in the garden and the bramble in the wood on which the adults fed are still there but the butterfly is only a memory.

The current list of 23 species is quite a good one. It represents 42 per cent of the British butterfly fauna. However the numbers of each species present are generally very small. The records show that the change from short grass to long grass with flowers benefits many species, whereas the change from grassland to woodland caused the virtual disappearance of grassland species. This was not surprising since cow parsley virtually replaced grass and very few sunny spots remained. If I did not manage the grassland in the Mere Enclosure it, too, would turn into a wood and also lose most of its butterflies. In recent years the area of grassland has been gradually reduced by the widening of the hedges and the invasion of blackthorn suckers. As mentioned above, I suspect that the recent declines in meadow browns and gatekeepers (Table 7 and Figure 10) have been due to this. I have little doubt that the greatly increased amount of garlic mustard maintains the small but steady population of green-veined whites and orange-tips, and that the patches of nettles in warm spots of the Mere Enclosure maintain at least some of the peacocks and small tortoiseshells; I find their caterpillars quite frequently. The grassland which contains a fairly wide range of grasses – perennial rye-grass, Yorkshire fog, meadow foxtail, cock's-foot, meadow barley, false oat-grass, creeping bent and red fescue – supports viable populations of browns and skippers. Plenty of ivy in our garden provides food for the second brood of the holly blue, but I do not know if our few hollies provide for the much smaller first brood. I have yet to discover whether our commas feed on hops or nettles – we provide both.

As noted above, annual fluctuations must be due to differences in weather, predation and parasitism, and possibly to unobserved changes in the different habitats as well as the large obvious ones. The numbers of migrants reflect conditions overseas. Those of painted ladies in Britain in 1996 were quite exceptional, and were reflected by the numbers I recorded in the Mere Enclosure that year. I had warning of what might happen because on 7th June I was doing a survey of the dragonflies on Dungeness for the RSPB. As I walked across the great expanse of shimmering shingle, thousands of painted ladies came flying in quite low to the ground and all heading north. They were accompanied by thousands of silver Y moths. It was an unforgettable sight. Sure enough a painted lady reached the Mere Enclosure two days later on 9th June. On 23rd June I saw two or three near Braemar in the Scottish Highlands. On 4th August I counted no fewer than 21 painted ladies feeding on thistles round the Mere. These were probably the offspring of migrants which had arrived earlier in the year. Some stayed until 16th August and I saw a single one on 5th September. As on Dungeness, the painted ladies were accompanied by many silver Ys.

Apart from the 24 species of butterfly which I have recorded in our small reserve, there are ten other species which breed or have bred recently within 20 miles of us, but there is little chance that they will colonize the reserve. We cannot provide the foodplants of the swallowtail, green hairstreak or chalkhill blue, for example, nor would it be easy to grow honeysuckle in the wood for white admirals. We do have plenty of blackthorn, the foodplant of brown and black hairstreaks but, as Jeremy Thomas has shown, these species are unable to disperse over open country and are very unlikely to reach us unaided. Each year I look up into our oaks to see if purple hairstreaks have found them. So far they have not but perhaps some day they will. Meanwhile we must make do with our present 23 species.

As a biologist I have been fascinated by the changes in the populations of our butterflies but the aesthetic pleasures I get from just seeing them around are every bit as important. Butterflies also make an increasingly valued link with the past. For example, I vividly remember seeing my first comma. It was on 5th September 1933 in Sussex, where I spent my childhood. The books said that the species was virtually confined to Hereford and Worcester. Actually it had begun to extend its range some time before, but I did not know that and so I was bowled over by seeing one sunning itself in our vegetable garden. Later I saw many more and I used to spend a lot of my school summer holidays looking for their caterpillars on the hops which were grown nearby. I found many of them, as well as the larvae of the pale tussock moth, which look like the heads of toothbrushes come alive. Locally these caterpillars were called 'hop dogs'. Today hops get deluged with insecticides and fungicides and hop gardens are virtually devoid of interesting insects.

Living near the coast in Sussex I often used to see migrant butterflies – painted ladies frequently and clouded yellows quite often. Once I was walking under the chalk cliffs known as the Seven Sisters. Huge numbers of white butterflies and small tortoiseshells were coming over the sea, but there was a strong offshore wind blowing and when they reached the cliff tops they were blown back, with the result that a great swirling circular cloud of butterflies accumulated on the seaward side of the cliff. It was an unforgettable sight. On another occasion I was walking along the top of Beachy Head when I saw a Bath white fly in from the sea. This was a most unusual event since, until 1945, very few Bath whites had ever been seen in England. I tried to catch it, but to my great disappointment it flew back over the cliff top and out to sea. However I was able to keep my eye on it and eventually it returned and I caught it. Today I would have released it, even though it had no hope of colonizing England, but in those pre-war days collecting seemed harmless enough. Whenever I see migrant butterflies in Swavesey today I am reminded of watching them in my childhood in Sussex.

Each year I look out for 'the first of the year' whether it is an orange-tip or a meadow brown. Seeing the first brimstone of the year is perhaps the most delightful experience: the brimstone is a harbinger of spring even if, like the proverbial first swallow, one may have to wait some time before real spring arrives.

## E. *Dragonflies*

Before constructing the Mere I had recorded seven species of dragonfly (azure, blue-tailed and common blue damselflies, southern and migrant hawkers, black-tailed skimmer and common darter) as visitors to our garden and plantations. Two, the azure and blue-tailed damselflies, bred in the little pond which I had dug in the Old Enclosure in 1968. More might have bred but for its huge population of newts. Nineteen species of dragonfly breed within five miles of the Mere in dikes, ditches, gravel pits and ponds and in the River Ouse. I had high hopes that many of them would colonize the Mere if it provided suitable habitats for them. My hopes were realised remarkably quickly: no fewer than 13 species appeared on the Mere in 1984 only a few months after it had been dug. They were five damselfly species – large red, azure, common blue, blue-tailed and emerald damselflies – and eight true dragonflies – migrant and brown hawker, emperor, four-spotted and broad-bodied chaser, black-tailed skimmer and common and ruddy darter. In later years I recorded a further six species: banded demoiselle, red-eyed and variable damselfly, hairy dragonfly, southern hawker and scarce chaser. In eleven years 19 of the 34 species which breed regularly in England (56 per cent) had found their way to the Mere. This showed how good dragonflies are at dispersing and finding new potential habitats, but it does not show whether the Mere actually provides suitable habitat for them to breed in.

Proof of breeding for dragonflies is much easier to obtain then proof of breeding for butterflies. This is because butterfly caterpillars are often very difficult to find while dragonfly larvae are restricted to water-bodies, and, when they change from larva to adult, they leave their last larval skins (exuviae) on the plants on which they have emerged. The exuviae of most species can be identified fairly easily as can well-grown larvae collected from the water. Recently emerged dragonflies fly quickly away from water to avoid capture as potential prey or mates. Therefore if you find recently emerged (teneral) dragonflies by water that also virtually proves breeding, even if you do not find their exuviae.

During the years since the Mere was constructed I have counted, whenever possible, adult dragonflies, male and female, teneral, immature and mature on transects round the Mere (c.95 yards) and in the surrounding grassland edged by hedges and bushes (c.195 yards), and have concentrated on the middles of fine days so that my observations on mature males would be comparable. I have also recorded finds of exuviae. The following account of how the Mere was colonized and how its populations changed is based on these observations. Needless to say there were times when I had to be away and so there are gaps in the information but none of my absences were very long.

Table 8 gives the details of how dragonflies colonized the Mere. It shows that of the 13 species which discovered the Mere in 1984, seven species at once bred successfully: five species which take a year to develop (azure, blue-tailed, common blue and emerald damselflies and common darter) emerged the following year, and two species which take two years to develop (emperor and black-tailed skimmer) emerged two years later. The brown hawker which appeared in 1984 and the southern hawker which appeared in 1985 also take two years to develop. They were not seen ovipositing until 1986: offspring emerged in 1988. This suggests that the Mere did not appear to be a suitable habitat to them in its first two years but did in its third. The larvae of the four-

Table 8    The colonization of the Mere by dragonflies in order of appearance, 1984–95

| Species | First year adult observed | First year oviposition observed* | First year breeding proven | Total number of years observed (1984–2000) |
|---|---|---|---|---|
| Azure Damselfly *Coenagrion puella* | 1984 June | 1984 | 1985 | 17 |
| Common Blue Damselfly *Enallagma cyathigerum* | 1984 June | 1984 | 1985 | 17 |
| Blue-tailed Damselfly *Ischnura elegans* | 1984 June | 1984 | 1985 | 17 |
| Broad-bodied Chaser *Libellula depressa* | 1984 June | | 1991 | 13 |
| Black-tailed Skimmer *Orthetrum cancellatum* | 1984 June | 1984 | 1986 | 7 |
| Emperor Dragonfly *Anax imperator* | 1984 July | 1984 | 1986 | 16 |
| Large Red Damselfly *Pyrrhosoma nymphula* | 1984 July | | 1988 | 15 |
| Common Darter *Sympetrum striolatum* | 1984 July | 1984 | 1985 | 17 |
| Brown Hawker *Aeshna grandis* | 1984 August | 1986 | 1988 | 17 |
| Four-spotted Chaser *Libellula quadrimaculata* | 1984 August | | 1989 | 17 |
| Ruddy Darter *Sympetrum sanguineum* | 1984 August | 1985 | 1992 | 15 |
| Emerald Damselfly *Lestes sponsa* | 1984 August | 1984 | 1985 | 17 |
| Migrant Hawker *Aeshna mixta* | 1984 September | | 1991 | 17 |
| Red-eyed Damselfly *Erythromma naias* | 1985 June | | | 3 |
| Southern Hawker *Aeshna cyanea* | 1985 August | 1986 | 1988 | 16 |
| Hairy Dragonfly *Brachytron pratense* | 1987 May | | 1988 | 12 |
| Banded Demoiselle *Calopteryx splendens* | 1992 June | | | 6 |
| Variable Damselfly *Coenagrion pulchellum* | 1993 June | | | 3 |
| Scarce Chaser *Libellula fulva* | 1995 June | | | 2 |

Notes:  (i)  *In column 2, the year is shown only when oviposition was observed before breeding was proven.
         (ii)   In some years only females of some species were observed.

spotted and broad-bodied chasers and the ruddy darter all require detritus in which to breed, so it was not surprising that they did not succeed in breeding for several years after they first appeared on the Mere. The ruddy darter (Plate 35) is essentially a marsh species. It was seen ovipositing as early as 1985, but I found no proof of successful breeding until seven years later in 1992.

I was particularly pleased that the Mere was colonized by the large red damselfly and the hairy dragonfly (Plate 36). These two species are unusual in belonging to genera which are confined to the Palaearctic Region. In fact the hairy dragonfly, the only species in its genus, is virtually confined to Europe. Both species used to be widely distributed in East Anglia, but since the Second World War both have declined, and in Cambridgeshire are confined to few localities.

The large red damselfly is one of our two red damselflies. Like the orange-tip butterfly it is a harbinger of spring. In England it appears in April and is nearly always the first to announce that the 'Dragon Season' has begun. I like it for another reason as it reminds me of some other red damselflies which it super-ficially resembles. These are the *Megalagrion* damselflies of the Hawaiian Islands. This one genus is found in a greater variety of habitats than any other I know. Different species live in ponds, streams, waterfalls, the little pools in the axils of epiphytic *Astelia*, and one species even has a terrestrial larva. When studying species of *Megalagrion* some years ago I was often reminded of our red damselfly and spring days at home, and when I see red damselflies by the Mere I am reminded of those extraordinary Hawaiian insects, which rival Darwin's finches in their wide radiation of forms within one genus.

The hairy dragonfly is a dark insect, the male with blue spots on the abdomen and the female with yellow. Usually one sees it flying up and down weed-filled ditches which are its favourite haunt. It is a hawker dragonfly and is unusual among hawkers in being a spring insect; the other hawkers are on the wing from mid-summer to mid-autumn. There are signs that the hairy dragonfly, which had declined so extensively, has been on the increase in recent years. For example, it has returned to Wicken Fen after a considerable absence. It has also returned to some ditches in the Swavesey Fens; the insects which found the Mere probably came from them.

The very striking broad-bodied chaser and the magnificent great blue emperor are also particularly welcome inhabitants of the Mere because until recently they were scarce in Cambridgeshire. The extremely aggressive broad-bodied chaser is well known to the owners of garden and farm ponds in southern England, where it is very common. The emperor is also familiar to visitors to large ponds and gravel pits in the south of England. It has a very wide distribution in the world, occurring over much of Europe, Asia, Africa and Madagascar. When we first came to Swavesey I very rarely saw either broad-bodied chasers or emperors. They are now notably more abundant.

**Table 9    Changes in dragonfly populations at the Mere, 1984–2000**

| Species | Largest number of males observed on transects by water in the year indicated (Transect length c.95m) | | | | | | | | | | | | | | | | | Number expected if Highest Steady Density were reached (see page 77) |
|---|---|---|---|---|---|---|---|---|---|---|---|---|---|---|---|---|---|---|
| | 1984 | 1985 | 1986 | 1987 | 1988 | 1989 | 1990 | 1991 | 1992 | 1993 | 1994 | 1995 | 1996 | 1997 | 1998 | 1999 | 2000 | |
| Banded Demoiselle *Calopteryx splendens* | | | | | | | | | 1 | | | | 1 | 1 | | 1 | | |
| Emerald Damselfly *Lestes sponsa* | | 5 | 2 | **14** | 12 | 5 | 5 | 10 | 9 | 5 | 13 | 9 | 5 | 6 | 5 | 8 | 6 | 99 |
| Large Red Damselfly *Pyrrhosoma nymphula* | 1 | | | 1 | 1 | 1 | 1 | | 1 | 2 | 3 | 4 | 1 | 1 | 3 | **5** | **5** | 27 |
| Red-eyed Damselfly *Erythromma najas* | | 1 | | | | | 1 | | | | | | 1 | | | | | |
| Azure Damselfly *Coenagrion puella* | 11 | 27 | 6 | 40 | 21 | 24 | 40 | 25 | 27 | **113** | 101 | 48 | 29 | 29 | 33 | 69 | 54 | 135 |
| Variable Damselfly *Coenagrion pulchellum* | | | | | | | | | | 1 | 1 | | 1 | | | | | |
| Common Blue Damselfly *Enallagma cyathigerum* | 4 | 20 | 16 | 15 | **30** | 14 | 9 | 1 | 2 | 5 | 9 | 8 | 9 | 6 | 21 | 4 | 9 | |
| Blue-tailed Damselfly *Ischnura elegans* | 11 | 27 | 6 | 19 | **30** | 10 | 11 | 7 | 5 | 5 | 4 | 3 | 7 | 5 | 8 | 13 | 7 | 27 |
| Migrant Hawker *Aeshna mixta* | 1 | 1 | 1 | 1 | 1 | 3 | 4 | 2 | 2 | 2 | 2 | 1 | 2 | 3 | 2 | **5** | 2 | |
| Southern Hawker *Aeshna cyanea* | | 1 | 2 | 2 | 1 | 1 | 1 | 1 | 1 | 2 | 1 | 1 | 1 | 1 | 1 | 1 | 2 | 2 |
| Brown Hawker *Aeshna grandis* | 1 | 2 | 1 | 1 | 2 | | | 1 | 1 | 1 | 1 | 1 | 1 | 1 | 1 | 1 | 1 | 2 |
| Emperor Dragonfly *Anax imperator* | 1 | 1 | | 1 | 1 | | 1 | 1 | 1 | 1 | | 1 | 1 | 1 | 1 | 1 | 1 | 1 |
| Hairy Dragonfly *Brachytron pratense* | | | | | | | | | | 1 | 1 | | | 1 | 1 | 1 | 2 | 8 |
| Four-spotted Chaser *Libellula quadrimaculata* | 1 | 1 | | 1 | 1 | | 3 | 1 | | 3 | 1 | 2 | 1 | 1 | 1 | 2 | 1 | 14 |
| Broad-bodied Chaser *Libellula depressa* | 1 | | | 1 | | 1 | 1 | 1 | 1 | 1 | 1 | | 1 | 1 | 1 | | 1 | 8 |
| Black-tailed Skimmer *Orthetrum cancellatum* | 3 | | | | | 1 | | | | | | | 1 | 1 | | 1 | | 9 |
| Common Darter *Sympetrum striolatum* | 5 | 7 | 4 | 6 | 7 | 7 | 8 | 5 | 6 | 9 | 7 | 7 | 7 | 6 | 9 | **13** | 5 | 12 |
| Ruddy Darter *Sympetrum sanguineum* | 1 | 2 | 1 | | 2 | | 1 | 3 | 4 | 4 | 5 | 6 | 4 | 6 | 3 | 6 | **7** | 14 |

Note:  Peak numbers of more numerous species are shown in bold.

Certainly in the case of the emperor there is evidence that the species is pushing northwards. It is a possible indicator of climate change.

The information in Table 8 shows how different species of dragonfly colonized the Mere between 1984 and 1995 but it gives no indication of the numbers of each species involved nor how populations changed over the years. Table 9 attempts to provide this information, however it needs some explanation. I have had neither the time nor the inclination to do exhaustive studies on the populations of dragonflies on the Mere; as with butterflies I have made do with transects. As explained above (page 63), if one makes transects of mature male dragonflies by water about mid-day on fine days one gets results which are comparable, and so I have used this measure as a yardstick.

While populations are quite low the number of males present probably provides a reasonable index of the population as a whole, but as numbers increase territorial behaviour makes this impossible because the number of surplus insects expelled from the pond is not known. I have made numerous studies on population density of dragonflies and have found that the population density of each species very rarely exceeds a certain value. What happens is this: as numbers increase the territory size of each male decreases, but a level is reached when the territories can no longer be compressed. This level is known as the highest steady density of the species. It is expressed as the number of male dragonflies per 100 metres of the edge of the habitat (i.e. of the water-body). It varies between species: it is much larger for small species with small territories than it is for large species with large territories. It is only a rough measure but it is useful in assessing the value of a particular locality for a particular species.

The numbers expected by the Mere if the highest steady density were to be reached are shown in Table 9. Of the species occurring in good numbers, only in three species – the azure and blue-tailed damselflies and the common darter – did the numbers approximate to the highest steady density. For all the other species the figures are considerably below their highest steady density. In some cases this may be because parts of the Mere was unsuitable for them, in others it may mean that the total populations were actually very small because few insects emerged that season.

The highest steady density of the very large hawker dragonflies is so low that, however many insects may have emerged from the pond, I could never hope to see more than one or two territorial males. Observations on other ponds, including some much smaller than the Mere, show that large numbers of hawker dragonflies may emerge from them, even if they are never patrolled over by more than one territorial male. Therefore records of only one or two hawker dragonflies on the Swavesey Mere do not necessarily mean that it supports very small populations of these insects. However, counts of exuviae did show that rather few individuals of each of the hawker species actually emerged from it. It appears that the Mere is an acceptable habitat for these species, but it is not a particularly good one.

The information in Table 9 suggests that there have been both trends and fluctuations in the dragonfly populations of the Mere. Once established the populations of fourteen species have been maintained, but while the ruddy darter has probably increased, the common blue and the blue-tailed damselflies have declined. The pioneer species, the black-tailed skimmer, soon became extinct as a breeding species. The azure damselfly population has fluctuated a great deal. Doubtless this and other species were affected by weather, food supply and predators. The gradual development of the Mere must have affected both food and predators. There were no fish in it until 1997 but there have been formidable aquatic predators – newts, the great water beetle (both larva and adult) and alder fly larvae among them. Despite the distance of the Mere from the nearest farmyard and before the extraordinary recent decline of the species, most years many house sparrows used to go to the edge of the Mere to catch emerging damselflies as they appeared. I have seen yellow wagtails preying on them too. For all these reasons it is not surprising that populations fluctuate from year to year.

Four species have visited the Mere but have left no descendants there. It is not surprising that the banded demoiselle and the scarce chaser have not bred as they are virtually confined to rivers. On at least one occasion a banded demoiselle arrived while I was at the pond. It flew round it and then away as if it knew that the Mere was not a suitable habitat for it. There are probably not enough pond weeds and water-lilies in the Mere for the red-eyed damselfly to colonize it. This species lays its eggs in the floating leaves of these plants and the males perch on them. The narrow eastern edge of the Mere is full of emergent plants and looks suitable for the variable damselfly but perhaps there is not enough of this habitat to maintain it.

All 19 dragonfly species which breed within five miles of the Mere have visited it. A twentieth species, the white-legged damselfly, breeds in small numbers in the Great Ouse about ten miles away. It is most unlikely to visit Swavesey and, if it did, the static water of the Mere would not be suitable for its larvae. I must be content with having attracted 19 species to the Mere and in having provided a breeding habitat for 15 of them. My main interest has been to see how its dragonfly population has developed – partly for its own sake but partly also because I could compare it with other ponds which I had studied elsewhere without the advantage of living close beside them.

I have also used the Mere to study particular problems about dragonfly behaviour. I have always been interested in the ways that the behaviour of dragonflies affects their numbers and hence in territorial behaviour. There is general agreement that territory in dragonflies is mainly concerned with repro-duction – in enabling males to mate successfully with as many females as possible. By marking individual dragonflies with coloured paint and releasing them I have been able to follow the movements of individual insects. Some return to the same spot each day but others set up their territories in different places. It is not obvious whether males defend particular localities, or whether they merely defend the area round them, wherever that may be. I carried out experiments with common darters by the Mere to see which they did. Common darters spend little time patrolling their territories, instead they perch on the ground or on a plant and make sallies at any intruding males which they see. They like to perch on warm surfaces and in the autumn I noticed that they particularly liked to perch on the trug or shallow garden basket I used for carrying equipment. I put this in the territory of a male common darter by the Mere and it obligingly perched on it. I then picked up the trug with the dragonfly still on it and very slowly and carefully took it to the other end of the Mere and put it down. The dragonfly remained but when another male appeared it attacked it before returning to the trug in its new position. I repeated the experiment several times with the same results. It seems clear that a common darter's territory is the ground and airspace round it not a particular locality.

Like those of the common darter, populations of adult male azure damselflies appear to be controlled by the insects themselves but, unlike male common darters, male azure damselflies never seem to attack and fight with other males of their own species. I spent much time sitting by the Mere watching azure damselflies and recording what they did. My observations confirmed that, although they never attacked other dragonflies of their own or other species, when a male approached them they would lift their wings in a threat display, like many other species. I put their ability to control their numbers to the test by carrying out experiments on small ponds at Woodwalton Fen. I added or subtracted individuals to or from populations whose size I had measured: whichever I did the population remained roughly the same size. In 'addition experiments', surplus insects left; in 'subtraction experiments', gaps were filled by males lurking on the edge of the ponds awaiting the opportunity to fill them. Presumably the threat displays are enough to maintain a limit to population densities. This is partly the case in many animals, but in most it is backed up by a real threat of force.

The large number of species occurring on the Mere has provided the opportunity to study the interaction of males of different species. The seasons of the brown, southern and migrant hawkers overlap. Males of each species vigorously chase out other males of their own species, but they also frequently chase males of other species. I have seen brown hawkers driven out by southern and migrant hawkers; southern hawkers driven out by brown and migrant hawkers; and, most commonly, migrant hawkers driven out by brown and southern hawkers – not surprisingly, since the migrant hawker is much smaller than the other two species. Nevertheless, the fact that it can drive out larger species shows that some advantage of familiarity with the territory or of sexual drive can sometimes overcome the disadvantage of small size. Fighting to defend a territory in order to maximize the chances of successful mating clearly has survival value. This 'explains' the fights between males of the same species. However, the value of fighting males of other species is not at all clear. My observations at the Mere show that it uses up a great deal of energy and sometimes causes injury to the insects involved. In a recent paper I suggested that its value to each male insect is to provide practice in the fighting skills needed to defend its territory from males of its own species.

I am currently studying the behaviour of dragonflies when they are away from the water in the Mere meadow and on its hedges. Although much is known about dragonfly behaviour and ecology there is much to be discovered – particularly if you are lucky like me in having dragonflies living on your doorstep.

## F. *The unrecorded majority*

My aim in recording the dragonflies, butterflies, birds and other vertebrates on my reserve was to assess how their populations changed as their habitats developed. These particular animals were chosen for the purpose because they are conspicuous and easy to identify in the field. As a result of studying them intensively other groups of animals were neglected. This section is about these neglected, unrecorded species on my reserve. What groups of animals did they belong to and what sort of number of species occurred in the reserve unidentified?

Observations showed me that examples of most of the main taxonomic groups which were likely to occur in the reserve did so. Of the groups not studied intensively I observed Mollusca (snails and slugs), Annelida (earthworms), Nematoda (roundworms), Arachnida (spiders and harvestmen), Crustacea (woodlice), Myriapoda (millipedes and centipedes), Collembola (springtails), Ephemeroptera (mayflies), Orthoptera (crickets and grasshoppers), Dermaptera (earwigs), Hemiptera (bugs), Thysanoptera (thrips), Raphidioptera (snake flies), Megaloptera (alder flies), Neuroptera (lace-wings), Mecoptera (scorpion flies), Trichoptera (caddis-flies), Lepidoptera (species other than butterflies, i.e. moths), Coleoptera (beetles), Diptera (two-winged flies) and Hymenoptera (sawflies, wasps, ants and bees).

To get some idea of the number of species which occur on my reserve but which have not been identified, it is helpful to make comparisons with Wicken Fen, 13 miles to the east of Swavesey, with which I have had a long association. Its most famous habitat is Sedge Fen which has no equivalent on my reserve. However, woodland, scrub, grassland, marsh and water do occur on both reserves. If we compare the number of regularly breeding species in the groups which have been studied intensively at both reserves we find the following:

Wicken Fen has 67 regularly breeding bird species, my reserve 20.
Wicken Fen has 24 regularly breeding butterfly species, my reserve 23.
Wicken Fen has 17 regularly breeding dragonfly species, my reserve 14.

All species which breed at Swavesey also breed at Wicken Fen. These figures show that despite the fact that Wicken Fen is about 500 times as large as my reserve and has a far greater wealth of habitats, the number of species in the three well recorded groups at Swavesey compare quite favourably with the equivalent species at Wicken.

The fauna at Wicken Fen has been studied for over 150 years. Over 4,500 species have been recorded there. These include over 1000 moths, over 1100 beetles, over 1800 two-winged flies and over 600 species of Hymenoptera. Observations in a wide range of habitats show that birds, butterflies and dragonflies do not occur on their own. They are accompanied by many other species – not only those on which they depend for food and shelter, but on numerous other species as well. All complex ecosystems like woods and ponds contain very many species. Therefore, if my reserve shares over 50 per cent of its well-recorded species with Wicken Fen, we can be sure that it supports at least hundreds of unidentified species in addition to the 200 or so which I have identified so far. The unrecorded species must far outnumber the recorded ones.

Among the species in the poorly recorded groups it is particularly enjoyable to meet those one has not seen before and which have probably arrived as the result of conservation management. For years we have had the little green oak bush-cricket in our garden; sometimes it enters the house. However I never saw a grasshopper until the lesser marsh grasshopper colonized the grass edges and banks round the Mere. This is quite a local species confined to the south-east parts of England. It is now common with us. Bumble-bees have also done well in the Mere Enclosure and have undoubtedly increased since we provided them with more suitable habitat. Scorpion flies have also increased, presumably because there is now more insect carrion for them to feed on. We see more hornets than we used to, but whether this is due to the new habitats or to a series of warm summers we do not know.

Providing aquatic habitats has enormously increased the biodiversity of the area. It was amazing how quickly aquatic insects discovered the Mere. I have already described how most of the dragonflies which now breed in it arrived during the summer following its construction. These were joined by a wonderful assortment of other aquatic insects. On the surface there were soon whirligig beetles, water stick-insects and water skaters. Beneath the surface large numbers of water-boatmen and back-swimmers could be seen. They were joined by several other species of water beetles including the impressively large great water beetle and the larvae of mayflies, alder flies, caddis-flies and china-mark moths. The introduced water snails have thrived. It is encouraging what an unpolluted pond can provide.

A special pleasure for the naturalist is to find an animal of a group which has hitherto been totally unfamiliar to him or her. I had this experience when I found an adult snake fly. This is a weird looking insect with membranous wings like those of an alder fly and an elongated neck, the predacious larvae of which live under bark. It is and looks like a primitive form of insect.

Other insects are a special delight because of their sheer beauty, especially those which are large enough to be observed without the aid of a lens or binoculars. I remember especially red underwing moths settled on a warm wall, a freshly emerged poplar hawk-moth in the Old Enclosure, a recently emerged lime hawk-moth perched at the base of one of our big lime trees and an elephant hawk-moth feeding on honeysuckle at dusk.

Many lifetimes could be spent studying the huge range of creatures which occur in this small bit of Cambridgeshire whose land use has been changed from agriculture to nature conservation. This thought provokes more general ones which are the subject of the second part of this book.

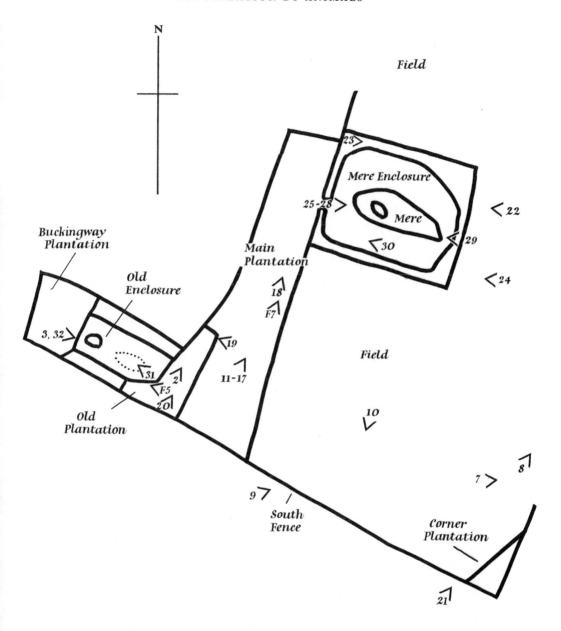

Figure 12: Sketch map of the reserve showing position from which photographs were taken. The base of each arrow marks where the photographer stood and the direction of the arrow indicates the orientation. All numbers refer to colour plates except those prefixed F which refer to black and white photographs in the text.

Plate 1 Aerial photograph taken on 29th July 1988 showing past and present features of our field.

Note indications of pre-Enclosure strips in the northern part of our field, which was later called Boxworth End Close. The most noticeable strip in the middle was a trackway to the large open-field system to the east. Two depressions in the southern part of our field (indicated by arrows), which was called Braziers Close, probably indicate earlier subdivisions. The hedge line between Boxworth End Close and Braziers Close can be seen.

By 1988 our woods had been planted and the Mere dug. They are clearly visible. At this stage the older and hence taller trees of the Buckingway and Old Plantations stand out from those of the Main Plantation, the Mere Shelter Belts and Corner Plantation as do the poplars by the Public Drain. For details see Figure 6. Neighbouring gardens and the disused orchard to the west of our field (see Figure 3) provide additional habitats for woodland birds (see p. 57).

Plate 2 The Old Enclosure in 1980. Until the 1990s there was often standing water in the hollow as shown in this photograph. The tree on the far edge of the pond is the one healthy alder in the reserve. On 6th November 1989 I put up a woodcock from this spot.

Plate 3 The Lombardy poplars in the Old Enclosure in 1973. These trees grew so well that they had to be felled in 1991 as a safety measure. Note the large bramble bush in the foreground which was much favoured by gatekeeper butterflies.

Plate 4  Meadow brown (*Maniola jurtina*). A butterfly which increased when grassland was enclosed and no longer grazed but declined as plantations matured.

Plate 5  Ringlet (*Aphantopus hyperantus*). Another brown butterfly which increased when grassland was enclosed. It appears to depend upon damper, more shaded places than the meadow brown and is much less common on the reserve.

Plate 6  Gatekeeper (*Pyronia tithonus*). The commonest butterfly on the reserve for many years. It needs both grassland and bushes. For a time it survived in the Old Enclosure after the Old Enclosure became too shaded for meadow browns.

Plate 7 Elms in the eastern hedge in July 1973. Note signs of Dutch Elm Disease in the two trees on the left. Note also the stumps of infected trees already felled to stem the spread of the disease.

Plate 8 Felling the last but one elm in the eastern hedge in 1979. The small elm at the end of the hedge was left as a memento. Note the sinuous outline of the hedge which, before the Enclosures, bordered the great open field to the east.

Plate 9 A large elm on the southern border of our field felled by the great gale of 3rd January 1976, photographed the following day. Dead and dying elms along the east hedge can be seen in the background.

Plate 10 A last look at the great elms on the southern border of our field, 1979. These dead trees were felled shortly after the photograph was taken. Two or three pollard stumps were left for a few years to provide nesting places for stock dove and little owl.

Plate 11  Planting the Main Plantation, December 1971 to January 1972. Peter is digging holes about six feet apart to receive the young oak trees. The soil is heavy clay and the holes often filled up with water.

Plate 12  The Old and Main Plantations, July 1973. The Old Plantation on the left is now eight years old and has reached the scrub stage. The ungrazed grass of the Main Plantation is taller than most of the oaks planted in it 19 months before.

Plate 13  The Main Plantation in late August 1975. Grass is still dominant; only trees like the hybrid black-poplars on the left, which were about five feet tall when planted, show above it. The plantation at this stage was not colonized by birds.

Plate 14  The Main Plantation in July 1976, a year of exceptional drought. Oaks are beginning to show above the grass. The elms in the foreground are suckers from trees in the south fence. The great elm has died of Dutch Elm Disease. Cattle have browsed well inside the post and wire fence.

Plate 15  The Main Plantation, Autumn 1978. Six years after planting, oaks and hazels are beginning to dominate the grass. The Poplars on Public Drain Number 12 are as high as the trees in the Old Plantation on the left.

Plate 16  The Main Plantation, Winter 1980. Much growth has occurred in the last two years. It has not yet been colonized by birds other than those which were already present in our garden and older plantations.

Plate 17  The Main Plantation, Autumn 1981. Ten years from planting, the plantation has become a young wood. The great elm has been dismembered. The plantation is just about to be colonized by new woodland birds.

Plate 18  Path inside the Main Plantation, Winter 1990. On the right is the hedge adjoining the fence on the edge of the field. Most of the trees seen here are oaks. There is more regeneration of young trees along the path than elsewhere in the Plantation.

Plate 19  Cowslips among cow parsley and cleavers in the Main Plantation, 1998. Where I have weeded round them, cowslips have thrived.

Plate 20  Daffodils in the Old Plantation, April 1990.
The leaves of dog's mercury and bluebells can also be seen. Wood anemones are present too. These species thrive in this part of the plantation only because cow parsley is weeded from it each year.

Plate 21  The Corner Plantation, April 1998. View of the plantation seventeen years after planting. Its hawthorn hedge is already in leaf. One of the hedgerow oaks can be seen growing in the eastern hedge in the distance.

Plate 22  Digging the Mere, 9th December 1983. Behind the JCB are the Lombardy poplars in the Main Plantation. Digging the Mere and placing turves on the clay banks took about two days altogether.

Plate 23  Shelter for the north side of the Mere, early Spring 1984. Immediate shelter is provided by the spoil dug from the Mere. Later it will be reinforced by the hedge and shelter belt planted just before this photograph was taken.

Plate 24  The eastern shelter belt of the Mere Enclosure, 1998. This shelter belt was planted on the site of the spinney which existed here in the 1830s. In the background are the Old and Main Plantations. Note the rooks' nests in the Old Plantation.

Plate 25  The Mere fills up, early Spring 1984. Spoil has been used to make a high bank on the north side and a low one on the south side. Aquatic plants were introduced later and the Mere was rapidly colonized by dragonflies.

Plate 26 The Mere, August 1984. Yellow flag irises and other species have been planted successfully. By the end of that month I had recorded no fewer than twelve species of dragonfly on the Mere.

Plate 27 The Mere, early Summer 1985. The Mere is less than two years old. Introduced emergent, submerged and floating water-plants have become established. The habitat is now ideal for dragonflies: fifteen species have been recorded and most of them are breeding.

Plate 28  The Mere, early February 1995. Until the recent periods of prolonged drought, the water level of the Mere has been at or near ground level by the end of the winter. The upturned boat under the elder bush (top left) is used for controlling bulrushes and reeds (Plate 29).

Plate 29  The Mere, February 1995, showing encroachment by bulrushes and reeds. It greatly increased after the exceptionally dry years of 1996 and 1997 and is now controlled by periodic cutting from the shore and by boat.

Plate 30  The Mere Meadow, April 1998.
The hawthorns and blackthorns of the south hedge of the Mere Enclosure are on the left. The large bramble bush on the right is much favoured by butterflies, and in 1997 sheltered a fox earth. Behind are oaks and poplars in the Main Plantation.

Plate 31  The Rookery in the Buckingway Plantation, April 1998.
The patch of nettles in the fore-ground is on the site of the Winter Mere.

Plate 32  The Rookery in the Old Plantation, April 1998.
The line of fruit trees is on the right. The dried-out pond is in the lower left-hand corner.

Plate 33  (Left)
Essex Skipper (*Thymelicus lineola*).
A local butterfly distinguished from the small skipper (*Thymelicus sylvestris*) by the black underside of the tip of the antenna. It occurs on the southern bank of the Mere.

Plate 34  (Centre)
Speckled Wood (*Pararge aegeria*).
An essentially woodland species. It was first seen in the Old Plantation in 1992, 27 years after planting. Two years later it was also observed in the Mere shelter belt which was then ten years old.

Plate 35  (Below left)
Ruddy Darter (*Sympetrum sanguineum*).
A local dragonfly which requires a marshy habitat. It appeared on the Mere eight months after it was dug but breeding was not proven until seven years later.

Plate 36  (Below right)
Hairy Dragonfly (*Brachytron pratense*).
A spring species which is virtually confined to Europe. It is very local in the British Isles. It became very rare in Cambridgeshire but in recent years has increased. Proof of breeding in the Mere was first obtained in 1988.

# PART TWO

## CHAPTER NINE

# MATTERS ARISING

S AWING wood and clipping hedges, mowing grass and cutting reeds can be quite hard work. Sometimes I have to admit my age and sit on the seat, which my son Peter gave me on my seventieth birthday. The seat is on the northern bank of the Mere, and from it I can look over the Mere and over the enclosure's hedge to the corner wood at the south-east corner of the field. 'Sometimes I sits and thinks and sometimes I just sits'. No, that is not quite true: a naturalist cannot just sit out of doors: I sit and observe. But when I do think I cannot help trying to relate my small-scale, practical experiences in my field in Swavesey to the wider conservation issues which I have had to deal with in my professional working life.

I have no doubt that the ideas which pervaded my working life as a conservation biologist influenced what I did in our field. My studies on habitat loss, whether on heathlands in Dorset or hedges in Cambridgeshire, made me long to put some variety back into our local impoverished landscape. I grew up in the Weald where woodland still exceeds 50 per cent of some of its parishes: the total absence of woodland in Swavesey made me long to have some on my doorstep, even though I knew a plantation was a very poor substitute for an ancient woodland. My studies on dragonflies and their conservation made me aware how many dragonfly species would benefit by digging new ponds.

Whilst my work influenced what I did to our field, what I did to our field influenced my official conservation work. It is so easy for environmentalists to say that farmers should do this and that, if they have no practical experience of doing the things which they recommend. Almost all conservation management takes time and money, and problems continually crop up. Initially it is difficult to decide on conservation objectives when they conflict, which they often do. Yet objectives must determine management. Once these have been decided upon, practical problems come thick and fast. Is the time spent in obtaining financial support for one's plantation worth the financial help obtained? It is highly desirable to use native stock when planting trees or making wildflower meadows but experience shows that this is often difficult to obtain. Is second-best, non-native stock better than nothing? Hedges make splendid habitats but it is several years before they are stockproof, therefore they need to be fenced initially, but fencing is expensive. What is the cheapest kind that will do the job effectively? Do you need permission to create a large pond? If you have to control its vegetation should you do this laboriously by hand or cheaply with machinery or herbicides?

I have had to face all these problems and many others in connection with my reserve. The experience has given me much sympathy for farmers who want to conserve wildlife but who are daunted by the genuinely difficult problems which confront them in carrying out effective action. Food production and conservation depend on the same resource – land. They are not easily reconciled when high productivity is required. Yet farmers and environmentalists must surely co-operate. The shared practice of conservation must help to bring us together constructively. I know it has in my case.

My small-scale endeavours at Swavesey reflect in miniature many of the huge problems which face all who are concerned with conservation at the beginning of the new millennium. I think something can be learnt from my experience so it is worth putting it in context – first, within that of the overall strategy for nature conservation in Britain.

# CHAPTER TEN

# NATURE CONSERVATION STRATEGY
# IN BRITAIN TODAY

F OR all practical purposes the objective of nature conservation is to promote biodiversity, that is to ensure the survival of as many species and ecosystems as possible. This objective should determine conservation strategy in Britain. Priorities have always had to be applied because resources for conservation have never been adequate. In practice, priority has had to be given to species and habitats which are threatened; in most cases threatened species are rare and their habitats are restricted. This has given the impression that conservationists are concerned only with rare species and establishing nature reserves to protect them. This impression is false. Ever since the start of the modern conservation movement in the 1950s and '60s, the strategy has been holistic: it has always involved two types of approach. On the one hand, special places have been protected by establishing nature reserves. On the other, attempts have been undertaken to make land outside nature reserves better for wildlife by influencing agriculture and forestry, through research and by seeking better control of pollution.

Since the 1960s rapid changes from traditional to intensive, highly productive agriculture have had very serious effects on wildlife. Essentially this is because modern agriculture in the short term is so efficient. Thanks to machinery and chemicals, weeds have been largely eliminated from arable and grass crops and with them the invertebrates which depended on both the crops and weeds. Birds and mammals which depended on the weeds and the insects have suffered accordingly. It seems very unlikely that nations like the United

Kingdom will return to low productivity, traditional agriculture on a large scale. Research – though not enough of it – is being done to discover whether new methods of organic or conventional farming can produce sustainable agriculture which is both highly productive and friendly to wildlife. Research on this topic is inevitably slow therefore quick results cannot be expected. For the foreseeable future, conservationists will have to operate in a world where most agricultural land will continue to support much less wildlife than it did 60 years ago. As a result, the small areas which have not yet been reclaimed, and thus still support much wildlife, are crucially important for conservation now because they provide the bases from which species can spread to the rest of agricultural land, if and when it can be made more suitable for them in the future. Many of these places have already been turned into nature reserves, but many more should be. Establishing nature reserves has never been more important than it is at this moment in history. Our generations have a huge responsibility to take this last opportunity to retain a sound base for conservation in the future.

Therefore it is very worrying that British programmes of nature reserve acquisition appear to be losing momentum. While no conservationist denies the need to protect outstanding sites today, the role of nature reserves has been downgraded. For example, in the 1990s expenditure on the acquisition of National Nature Reserves and on their management by Government conservation agencies was proportionally only about half what it was in the 1970s. As regards these agencies (English Nature, Scottish Natural Heritage and the Countryside Council for Wales), the change has been partly due to the party political dogma of some influential Conservative politicians. Whereas in the past it was considered desirable that the state should itself protect habitats in perpetuity on behalf of the people, in recent years public ownership of land has been considered at best unfortunately necessary and at worst undesirable. Other issues have also diminished the perceived value of nature reserves. For example, there is more scientific information about the vulnerability of small isolated habitats and their constituent species. The splendid aim to involve more people in conservation has put emphasis on conserving wildlife everywhere rather than on rare species in rare habitats. Reserves are expensive to manage and so when more are acquired there is less money available for other valuable conservation activities. Taking all these facts together it is not surprising that the national agencies acquire fewer nature reserves nowadays. There is much to be said for involving all competent conservation bodies in the business of protecting the natural heritage, but this should not be seen as an excuse greatly to reduce the number of reserves acquired by the agencies themselves. Many potential sites still exist and are threatened. They should be added to what the late Sir William Wilkinson, the last chairman of the Nature Conservancy Council, so rightly called 'the Jewels in the Crown'. The psychological value of maintaining an active nature reserve acquisition programme is considerable. State ownership of these special sites emphasises not only the importance of the sites themselves, but also the importance of conservation in the life of the nation. We should cultivate a pride in these wonderful places protected for all of us and for all future generations.

Many non-governmental organizations have also become less interested in acquiring new reserves in recent years because of the cost of their management. Therefore, the less well-endowed ones prefer to carry out less costly conservation activities such as organizing conferences, setting up information centres, providing interpretative leaflets and attempting to influence the managers of land.

The change in strategy which results in putting more resources into public relations and less into nature reserves fails to take account of the time in which we live. Our generations really are the last to have the opportunity to protect sites on a large scale. Soon most sites will no longer be available to be made nature reserves: they will have disappeared. So, while retaining our vision of valuing all wildlife, by far the most urgent task for *our* generations is to hold the fort by protecting existing outstanding habitats until the long, slow process of changing the management of agriculture allows wildlife to return to the whole countryside.

While top priority for government and voluntary bodies should go to establishing nature reserves, much can be done to conserve wildlife on farms which are managed intensively. As well as increasing the total amount of wildlife, conserving habitats on farms will provide stepping-stones, and sometimes even corridors, by which individual animals can move between nature reserves, thus making their populations more genetically viable. Under the conditions imposed by intensive agriculture, habitat conservation on farms can be achieved only by zoning – that is by dividing farms into two zones, one covering food production, the other managed for wildlife. On most farms the two zones already exist: first the fields producing food and second a smaller area made up of hedges, ditches, woods and ponds, which do not produce food, but whose function is or was to support food production indirectly. Most species on the modern farm are confined to these places and so they should be conserved as far as possible and increased by planting new hedges, shelter belts and wildflower meadows, making conservation headlands and by digging new ponds. The desirability of conserving and creating these habitats is accepted by a growing number of farmers and advice on what to do and how to do it is readily available from advisers, notably from the Farm Conservation Advisers of the Farming and Wildlife Advisory Group (FWAG). What percentage of land each farmer can devote to conservation will vary from farm to farm. It should not be less than 5 per cent; many farmers could manage considerably more without seriously reducing the profitability of the whole farm. What we ask each farmer to do on the small scale is what the nation does or should do on the large scale – to conserve the best habitats on his/her land. So long as we retain a long-term vision of devising systems by which nature conservation can be linked effectively with agriculture, forestry and urban life, the half-measures just outlined are the best and probably the only way we can hold the fort in these critical times. My reserve at Swavesey is a tiny contribution to the process. It is encouraging that so many species can be supported in so small an area.

How the conservation of plants and animals relates to the conservation of our own species is discussed in Chapter 13.

# CHAPTER ELEVEN

# THE PROS AND CONS OF CREATED HABITATS

AS mentioned in the previous chapter, farmers can further conservation on their land both by retaining and managing existing habitats and by creating or recreating new ones. To which activity should they give priority? Indeed, to what extent is it possible to recreate habitats which have been lost? This is an important question. It is often raised when a developer wishes to build on the site of an ancient wood, meadow or heath and maintains that this is acceptable if he creates new wildlife habitats in the vicinity. Is he right? Experience at Swavesey suggests answers.

My land contained no woodland, herb-rich grassland nor pond, so I had no choice. If I wanted to improve the land for wildlife I had to recreate habitats. How do they compare with well-established equivalents?

My wood consists largely of native tree species of British stock and, apart from the absence of large, old trees, it now looks superficially like the oak/ash woods which are native to the area. However, in its detail it is completely different. There is little dead wood. Its ground flora of cow parsley, ivy and cleavers bears no relation to the ground flora of dog's mercury, bluebells and primroses which must have originally existed on the site. I have planted these and other ancient woodland species in the wood but, as explained on pages 40–42, I can maintain them only by weeding out the cow parsley, ivy and cleavers when they threaten to take over. My wood is still very young, but much older plantations in Cambridgeshire have ground floras like mine and not like those of ancient woods in the county, such as Hayley Wood and

Knapwell Wood a few miles away. Therefore, as far as woodland is concerned, plantations are clearly not adequate substitutes for ancient woodland even if they consist of native trees. Farmers should give priority to conserving ancient woodland if they are lucky enough to own some, and developers should not pretend that plantations are adequate substitutes for ancient woodlands which they propose to destroy.

All this does not mean that plantations have no value for conservation. They are enormously better for wildlife than arable fields or grass leys. The trees themselves and the insects which feed on them are most valuable. If native trees are planted the bird and mammal fauna of plantations may become as rich as that of ancient forest. For many farmers this is sufficient reward for planting. I can sympathize as I have derived enormous pleasure from watching blackcaps and other birds colonize my little wood.

The situation with grassland is not very different from that of woodland. While I have succeeded in introducing some of the local species to my meadow, I have failed with others. Discussions with FWAG Farm Conservation Advisers show that my experience is typical: recreating flower-rich meadows is a difficult business. Again time is of the essence: it can take a very long time to impoverish the soil enough to prevent the few aggressive species from excluding the many less aggressive species. Even when a fairly suitable flora has been established, it seems to take a long time before it acquires its typical fauna. In general it is much more effective and very much cheaper to maintain existing grasslands than to recreate them. However flower-rich grasslands are now so rare in Britain that it is highly desirable to add to their number by trying to recreate them even if efforts are only partially successful.

Ponds are a different matter. They are naturally transient things, and therefore a new man-made pond is unlikely to differ much from a natural one. It is not surprising that my Mere has a fauna as good as any pond I know in Cambridgeshire. Ponds are one of the few habitats one can truly recreate. They have the additional advantage that few other habitats support so many species in such a small area.

To conclude, of the habitats which can be created on farmland, only ponds are equivalent to ancient well-established ones. Plantations and meadows sown with wild-flower seeds are well worth creating but their conservation value is considerably less then that of ancient woods and ancient meadows. These last are the places which should receive priority within the conservation zone of the modern farm.

# CHAPTER TWELVE

# CONSERVING THE UNKNOWN

MOST animal species are invertebrates. If farmers and others accept the value of biodiversity and the need to conserve it, how can this be done when so little is known about most invertebrates and so little time is left when we can be effective in their conservation? This question concerns all who establish nature reserves.

The scale of the problem is illustrated by what I have been able to do and what I have not been able to do with our reserve in Swavesey (see page 80), but it applies equally to large reserves and vast national parks. Swavesey is in Britain, an offshore island in the temperate zone; as such it has a relatively poor flora and fauna. Despite this fact, I am nowhere near being able to identify most of the species which live in the wood, the grassland and the Mere in my reserve. It is relatively easy to identify the flowering plants and the larger animals which feed by day – birds, butterflies and dragonflies – but the lower plants and most invertebrate animals are harder to find and often harder to identify and I simply have not had the time to do either. Had I been less busy I could have done more, but not much more. This situation exists in nearly all the world's protected areas, whether national parks or nature reserves. There are a few places, such as Wytham Wood in Oxfordshire, Wicken Fen in Cambridgeshire and Barro Colorado in Panama, that have been studied comprehensively by teams of people, but these reserves are highly exceptional. For most nature reserves, species lists consist of little more than lists of trees and other conspicuous plants and diurnal mammals and birds. Even in Britain, where the flora and fauna are relatively poor in species and well known and where the population-density of biologists and naturalists is well above average, the species lists for most reserves are very patchy.

The situation is not as hopeless as it looks. Each species, whether studied or not, depends upon a particular habitat. Therefore if we conserve examples of all the main habitat types in the world we will conserve most of the world's species, known and unknown. There is good evidence to show that this supposition is correct. In Britain the network of National Nature Reserves covers most of the main types of habitat. The reserves have been selected primarily on what can be observed easily – their vegetation. No attempt has been made to select nature reserves specifically for dragonflies and yet studies show that practically all the species of British dragonflies are protected in one or more NNRs. It is almost certain that most species of other groups of invertebrates are given some protection by the network of NNRs.

The degree of protection afforded will, of course, depend first on the rigour and comprehensiveness of the classification of habitat types, and secondly on the size of the reserves chosen, which will determine the viability of the populations which they contain. However, there is no doubt that, if every country conserved reasonably large areas of all its main habitat types, the vast majority of invertebrate species – both known and unknown – would be conserved and hence the world's biodiversity. Since it is impossible to recreate most of the world's habitats, no alternative strategy is available and there is no excuse for governments and conservation organizations to delay taking the measures necessary to protect examples of all their existing habitats. The present rate of habitat destruction throughout the world makes it vital that we protect these areas while we still can.

In the circumstances, all we *can* do for the thousands of unknown and little-known invertebrates is to conserve viable examples of the main habitat types as outlined above. Lack of knowledge must not be used as an excuse for inaction but as a spur to complete the establishment of networks of national parks and nature reserves while this is still possible. This strategy provides the bare minimum of what is required if we are to avoid mass extinctions. The more that farmers and foresters can back it up by protecting habitats on their land, the more effective it will be as a basis for conservation in the future.

# CHAPTER THIRTEEN

# CONSERVING WILDLIFE CONSERVES
# HUMANS

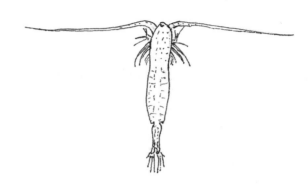

I N Chapter 10 it was assumed that the national objective for conservation
was to conserve as many species as possible. On the small scale this was the
tacit assumption behind all the practical decisions I made in establishing
and managing my nature reserve at Swavesey. However, even at Swavesey
experience shows that conserving one species can lead to the disappearance of
others. For example, reeds threatened to eliminate most of the other water
plants on the Mere, and so maintaining as many water plants as possible
involved controlling the reeds. While welcoming the arrival of the grass
snake we may be in danger of losing our newts. The maintenance of maximal
biodiversity in one place is clearly not a simple matter.

When we take a world perspective and include *Homo sapiens* among the
species to be conserved, the problem becomes even more complicated: we are
forced to consider what 'conserve as many species as possible' really means.
The human population is set to increase for another hundred years at least.
This will reduce the conservation options for most other species. To achieve
significant results in the future, measures to conserve wildlife will have to be
integrated more closely with measures to conserve people. We have reached a
stage in history where mankind has to ask very fundamental questions about
the total objectives of conservation.

It is not known what most people think about the relevance of wildlife
conservation to human welfare – they probably do not think about the
problem at all. When challenged about what should be done about wildlife,
opinions would range from 'couldn't care less' to passionate concern for all
living beings. But it is safe to assume that the vast majority would agree that

the welfare of the human species was paramount. It follows that the common ground on which all would agree about wildlife is that we should conserve those plants and animals on which human survival depends. The problem which arises from this utilitarian approach is that we do not know which species are crucial. All we know is that we are dependent on productive farm-land for food and on the oceans and forests for the functioning of the biosphere as well as for food and fibre. Therefore we depend on the species which make these ecosystems tick. They include very abundant species. In complex ecosystems like rainforests the whole system may also depend upon the survival of a few quite uncommon species in the food web. Clearly if humans wish to survive they must give top priority to ensuring that both abundant and uncommon 'key species' are not seriously damaged by pollution or excessive exploitation. As we learn more about the working of ecosystems we will be able to be more specific about which species are critical for human survival.

Most species are relatively rare and are probably not crucial for the survival of the systems of which they are part. For example, when the Tasmanian wolf became extinct the Tasmanian rainforests did not collapse. These rare species, which are not 'key species', may not be necessary for mankind's brute survival but it is becoming increasingly obvious that many of them will prove to be of inestimable benefit for future generations. They will provide new crops and new domestic animals and sources of drugs to combat diseases, not to speak of subjects for scientific research into our own nature and the functioning of our planet. Most would agree that this vast potential for the future should be maintained as far as possible. We cannot know which of these species will prove to be of special value in the future. A very large proportion would be conserved by conserving tropical rainforest where biodiversity is at its greatest.

There are also numerous species in different habitats throughout the world which enrich our lives by their beauty and interest. They may not be crucial for our survival or prove useful in the future, but most would agree that their survival is highly desirable so that future generations can enjoy them too. Their preservation is desirable in the same way as is the preservation of great works of art.

When finance for the future has to compete with finance for today, priorities have to be made. Clearly conservation for Man's survival must receive top priority. Fortunately the objective of conserving species which have potential value for future generations and the objective of conserving species of aesthetic and cultural value, are both largely subsumed in the requirements for our own survival – the conservation of the great ecosystems particularly the oceans and rainforests. Thus if we take effective action to conserve ourselves we should conserve most wildlife. And so it is that both those who confine their concern for plants and animals to utilitarian purposes and those who believe that mankind has an obligation to conserve all wildlife for its own sake can make common cause. Differences in philosophical objectives need not stand in the way of action. The possibility of conserving most of the species which exist today is still with us, but only if we act decisively and quickly to conserve ourselves. In the next chapter I shall discuss the obstacles which seem to prevent us from ensuring our own survival. I shall then suggest how these obstacles might be overcome.

# CHAPTER FOURTEEN

## CARE FOR THE FUTURE IN
## THE PRESENT

UNTIL recently I felt that I had achieved at Swavesey what I had intended. I had a plantation of maturing trees which had attracted a number of woodland birds and other animals. I had managed grassland so that it supported a rich butterfly fauna. The Mere had been colonized by a wealth of species, including all the dragonflies which could be expected. I felt in control and was optimistic about the future, but now something is happening which threatens what I have achieved and I have no control over it. Recently we have experienced the driest two-year period in England since records have been kept. In 1996 rainfall was only about 60 per cent of the average and in 1997, at a time when we expected water levels to be topped up, we had only about one and three-quarter inches in the period 1st January to 30th April. Effects which began to appear earlier intensified. Most of my birch trees died; several ashes and willows lost their tops. I lost some of my most treasured trees. For example, the large service-tree in the Buckingway enclosure died, as did the smaller one in the Main Plantation. Our largest beech, which Peter had brought with us as a little sapling when we moved from Dorset in 1960, also died in 1997. In the autumns of 1996 and 1997 the Mere dried out to such an extent that it was possible to walk across to the island from the south shore. Normally the Mere starts the spring completely full, but in 1997 and 1998 its level was about two and a half feet below the normal. Its low level allowed reeds to colonize parts which were once too deep for them. If the drought had continued, it might have dried out completely like many other ponds in the area.

These happenings were probably caused by what is often called global warming but which is better described as climate change. As predicted by experts on climate change, we have experienced different kinds of abnormally extreme weather in recent years. The droughts of the late nineties were followed by exceptional floods in 2000. November that year was the wettest since records began. Nearby, thousands of acres of the Ouse valley were under water. The Mere filled up to the brink and then overflowed, but of course the rain came too late to save the trees lost in the drought. No one can be absolutely certain that the droughts and the more recent floods are not just exceptional 'blips'. However, the records for the last decade suggest otherwise. If the climate is changing, the change is almost certainly due in part to human activities over which I have no control, except for the pollution I myself cause. The sad fact is this: individuals can do much to conserve wildlife in their gardens and on their farms but their endeavours are at the mercy of factors which only governments can control. If I had not been concerned already about such matters, exceptional drought and rainfall in Swavesey would have forced me to face the wider issue of governments' involvement with the environment.

If current predictions on global climate-change are correct, the implications for mankind and wildlife are horrendous: 200 million people will be endangered by flooding within the lifetime of people living today, and 50 million put at risk from starvation. Habitats will be altered so rapidly that many of their plants and animals will fail to adapt fast enough. Some species with less exacting requirements will be able to survive by moving, but frequently there will be no suitable places to which they can go. If the present trend continues we must expect that mass extinctions of plants and animals will follow.

Quite apart from the effects of global climate-change, rainforests continue to be destroyed on a vast scale by uncontrolled logging. In many places the rate of destruction has declined only because there is little forest left to destroy. The loss of forest plants and animals includes many which could have been very valuable to mankind in the future.

Over-fishing continues throughout the oceans. Even in sophisticated high-tech Western Europe no serious efforts are being made to control fishing under flags of convenience, a practice which further undermines the inadequate efforts of European governments to control the activities of their own fleets. Again and again fish quotas, which have been carefully worked out by scientists dedicated to the survival of the fishing industry, are watered down by politicians more concerned about fishermen's votes today than the long-term survival of the fishermen's livelihood.

When the potential dangers for mankind and the environment are so huge and far-reaching it seems extraordinary that sensible precautions are not taken to reduce them. The failure to act demonstrates that there must be formidable obstacles which prevent action. Progress clearly depends upon understanding the obstacles.

First we must ask whether there is anything fundamental about the human psyche which prevents people from undertaking measures to protect future generations. Most of human evolution occurred when mankind was a hunter gatherer: natural selection produced individuals who were adept at ensuring their own survival and that of their offspring. This was enough to safeguard the continuity of the species when environmental circumstances changed relatively slowly. However, there was no evolutionary mechanism through which mankind could become pre-adapted to deal with major environmental change in the future. In other words there is nothing innate in human nature which helps us to have a long-term concern for the future. Thousands of years attending to immediate problems may have made us believe that only the immediate matters: we are not helped by our cultural history any more than by our evolutionary history to take long-term views. Insofar as we base our current behaviour on past precedents, our culture may disincline us to consider the future seriously.

Modern Man is in a particularly poor position to understand problems which concern the living world because half the world's population now lives in towns and therefore has no day-to-day experience of its total dependence on nature. On the contrary, for millions the 'real world' seems to consist solely of business, the acquisition of goods, fashion and entertainment, the whole linked together and reinforced by increasingly efficient information systems. The living environment is seen as something outside day-to-day experience and hence peripheral. By most it is not seen as something we depend upon and therefore something which *must* be conserved.

The shortness of human life may further disincline us to consider the future: even if we appreciate the changes, we may think we shall be dead before environmental catastrophe catches up with us. Even our children and grand-children may escape, so there is no need to bother – at least not yet! We must conclude that mankind's evolutionary and cultural inheritance and the way he lives today give little support to taking the future seriously.

The 'do nothing' approach is further supported by lack of certainty. Experience has shown us repeatedly that we have failed to predict major events such as the onset of Aids or the collapse of the Asian 'tiger economies'. This makes many feel that we cannot predict anything and therefore there is no point in taking any action about the future. This attitude is clearly wrong: there are many things we can predict. For example we can predict with absolute certainty that if we destroy habitats we will exterminate the species dependent upon them. We can and should act on this certain prediction by conserving habitats. We cannot make such accurate predictions about climate change: we do not know enough to forecast the detailed effects of Man's pollution of the atmosphere. However, we can be certain that harmful effects will occur if pollution is not brought under control. In this case we have to act on probability. This is what we all do in our everyday lives. We decide whether or not it is safe to drive on an icy road or to walk down a particular street at night. But it is much harder for governments to take big decisions when so many people would be inconvenienced by them in the short term and wrong decisions might sometimes be made. If the decision not to act is likely to produce catastrophe, this risk must be accepted and precautionary action taken. Precautionary action is no more nor less than taking out an insurance policy.

Lack of certainty about the objectives of conservation, which were discussed in Chapter 13, also inhibits effective action. It is not surprising that the people who have the clearest grasp of conservation objectives are those most concerned with wildlife and hence its survival. They are the ecologists, naturalists and nature lovers in general who are watching what is happening to habitats and species, and are rightly alarmed at what they see. These people have a key role in modern society in getting the general public to face the facts and understand the environmental crisis of mankind and nature.

Unfortunately some nature lovers can be their own worst enemies. Some of those who have established close relationships with their pets and other animals and are incensed by the cruelties perpetrated against domestic and wild animals, equate animals with human beings. When taken to extremes this view discredits not only its proponents but the whole environmental movement and so makes it less likely that other less strident but more thoughtful voices will be heard. Unfortunately the media thrive on extremes and so give more prominence to such views than to balanced ones which could lead to effective action. Those who take extreme views about animals must come to terms with mankind's ecological position within his environment. Man has always had to kill plants and animals. The invention of agriculture greatly increased the number he killed, because not only did he kill the crop plants and domestic animals on which he depended, but also millions of individual wild creatures when he cleared the forest to grow crops. Once fields were established, weed control and drainage annually cause the deaths of many more organisms. When it was realised that mosquitoes were vectors of disease, the drainage of marshes killed not only the mosquitoes but also large numbers of other species both invertebrate and vertebrate. Modern Man may not wish to notice it but, whether omnivorous or vegetarian, he does have to kill to survive. Beliefs which ignore mankind's position in the global food web are untenable and undermine action.

Despite all the uncertainties and difficulties, large numbers of individual people show by their actions that they want to do what they can to protect the environment. As a result much waste is recycled, unleaded petrol is extensively used, and more new houses are heated with solar panels. That so many firms advertise their products and wrappings as 'environmentally friendly' shows how widespread is the wish not to damage the environment unnecessarily. Much can be done by individuals but the involvement of Government and hence politicians is essential. Unfortunately it is not easy for politicians in a parliamentary democracy to do what is required. They find it difficult to be too far ahead of the electorate. Some do recognize the need for robust action to save the environment, but they are aware that the means to achieve it would restrict their constituents' freedom to consume and to travel to some extent, and they are aware that that would be unpopular. It is true that some of the necessary environmental action would result in some jobs being lost; it needs great strength of character in an MP to acquiesce in the loss of jobs in his constituency. Of course, changing from polluting sources of energy to non-polluting ones like wind, wave and tidal power will demand new techniques and services which will create many new jobs, but until these opportunities are made available through government action, retaining old jobs will always be a more popular option for politicians in the short term. To get over this problem the public must convince politicians that they really do want government to do what is necessary for the environment even if it is at some inconvenience to themselves. The outcry about fuel prices in 2000 showed that we have a long way to go. The first requirement is to get the public fully aware of the true scale and urgency of the problem so that the majority demands action by Government through the normal political processes.

I believe that one of the chief obstacles in getting over the message is that we are dealing with what is essentially a new concept – the linking of immediate and future requirements in day-to-day political activity. Without a word which expresses this, thoughts and opinions will remain woolly and unfocused. I suggest that we can learn much from the success of coining the word 'biodiversity'. The concept behind this word is not new – conservation has always been about the value of the variety of life and the need to conserve it – but having a word which symbolizes the conservation objective has focused the problem and has made many people aware of it for the first time. Similarly there is nothing new in saying that we should plan for the future as well as for the present; many activities do that already. But what is needed is a term which defines the particular political activity which brings this about. It must cover the fusion of the two so-called 'real worlds': the familiar one of daily life, commerce, media, sport and entertainment, and the more fundamental one of the natural world and its life supporting systems. Until it is recognized that the two 'real worlds' are two sides of the same coin, we cannot expect true realism and progress in politics. Tentatively I suggest the new term 'Future Care' politics. Once this or a better term is in circulation, people will ask what it means and so will be introduced to the concept. Once convinced of the necessity for Future Care they will ask their MP what he or she is doing about it.

Unfortunately the ordinary political timetables in a parliamentary democracy stand in the way of taking the long-term approach which is essential for conserving mankind and its environment. The frequency of general elections forces MPs to take a short-term view, and even if one government embarks on a long-term policy there is no guarantee that the next government will follow it. These facts suggest that a bipartisan policy is necessary. In times of war politicians are forced by circumstances to forget differences and seek the common good. The future of the environment is also a matter of life and death, and so political parties should be able to put aside differences and achieve an effective long-term campaign on behalf of future generations. However, this could be a short-term arrangement only: if prolonged, it would stifle debate and no government would permit important decisions which were outside democratic control to be made for long. In the long run it would be far better for political parties to vie with each other in producing the best Future Care policies. Of course there is a possibility that a party will pander to selfish instincts and will say that its policy is to wait and see and not do anything about the future. This would be better than all parties ignoring the issue, because argument would keep it in the public eye. A prime minister, who was at least partly a statesman, would not find it difficult to pillory those who showed so little concern for their children and grandchildren. For the public, sick and tired of short-term politics and of politicians who lack vision, Future Care would stimulate exciting new discussions about goals and opportunities and would help restore faith in parliament.

Future Care would be applied in all fields, most crucially those dealing with resources such as fisheries, land-use planning, water supplies, the management of flood plains and coastal erosion, but also in transport. In the 1960s it was obvious that transport would increase. If Future Care had been adopted then, British Rail would not have been allowed to sell off its disused railways, as they (or roads built on their sites) were bound to be needed in the future. The present Government's transport policy would have been greatly helped by being able to take disused railways out of mothballs and rehabilitate them.

In every case the demands of the present would have to be weighed against the demands of the future. Special protocols would have to be achieved to ensure that future generations were properly represented on executive committees. Disputes between representatives of present and future might have to be regulated by new forms of law. I have no doubt that once the principle of automatically and seriously considering the future in all relevant activities became universal it would come to be seen as common sense. Arguments would not be about whether Future Care should exist but about the best ways of implementing it.

The problems described above are related principally to one rather small country in Northern Europe, but the survival of the environment and hence of mankind is a global problem and in an ideal world should be solved globally. However, in the absence of world government, environmental action has to be achieved by national governments. International agencies such as UNEP and IUCN can and do give excellent advice. The more efficient globalized firms take the future increasingly into account. Internationally agreed conventions have made some progress in limiting the excesses of the industrial world. One notable example suggests a way for the future: the Antarctic Treaty. This has resulted in nations putting territorial claims on that continent into abeyance and has kept unsuitable and potentially dangerous commercial enterprises at bay. It has encouraged international co-operation in scientific research for the benefit of the whole world.

If this can be done for the Antarctic it should not be too difficult to do it for the oceans which are outside territorial borders. At present the oceans are suffering from a deadly combination of largely uncontrolled high technology and a 'hunter-gatherer' philosophy totally out of keeping with the modern world. Failure to treat the oceans as the common heritage of mankind has led to the extinction of immensely important fish stocks and the consequent ruin of whole communities. Before worse befalls, the nations of the world must act so that the products of the oceans can be conserved and harvested instead of being slowly but surely destroyed through a Stone Age approach.

A strategy similar to that for the oceans should be adopted for the rain-forests. But here the problem is much harder because all the forests occur within the borders of self-governing nations. We all depend upon the rain-forests, so surely it is the duty of all nations, especially the affluent nations of Europe, North America and Asia, to give financial support to the efforts of impoverished Third World countries to maintain their forests.

Affluent countries, notably the USA, produce far more pollution per head of population than do other countries. They have a special duty to cut down on the amount of pollution they currently produce. The appalling disasters recently experienced in Mozambique and in parts of India are likely to increase unless the world's wealthy nations make much greater efforts to control pollution than they are doing at the start of the new millennium.

Pessimists might argue that it is useless to undertake far-reaching national environmental action in the absence of effective international action. This is a false proposition because all action helps. As the environmental crisis develops it will became obvious that proper action has to be taken. When this happens nations which have already begun to take appropriate measures will be at a distinct competitive advantage. Therefore we in Britain should not wait for others to catch up, we should put Future Care into operation as soon as possible.

By our actions we show that we are perfectly capable of acting for the future: we plan now for the future of both the national health service and education. Illogically and dangerously we fail to plan for the future of the environment, the condition of which will affect everything we do in the future. We have to ask: will the necessary action take place before the situation becomes acute – before huge numbers of species are exterminated and mankind's future is seriously in jeopardy? Past experience is not encouraging: effective action usually occurs only after obvious disasters. However the number of people concerned about the environment has increased greatly in recent years, so we may just succeed in getting action ahead of catastrophe. If this happens there will have been a fascinating sequence of events. The concern of numerous individual naturalists will have acted as a catalyst in conserving mankind, the species which unthinkingly nearly destroyed its own habitat. Thus there will be a direct connection between the experience of individual naturalists, such as that described in the first part of this book, with the conservation of our own species. Small things are inextricably mixed with big things, so it seems as I sit on my seat by the Mere watching dragonflies and grandchildren.

# CHAPTER FIFTEEN

# POSTSCRIPT

**M**UCH has happened since the manuscript of this book was first submitted to the publishers. Hugely significant international events have occurred which are likely to affect every environment in the world. Before trying to relate the topics in the second part of this book to the new circumstances, I shall bring up-to-date the changes which have occurred on my reserve – the little patch of ground from which I contemplate the larger scene.

The exceptional rainfall of 2000–2001 (see p. 99) appeared to have a marked effect on the vegetation of the reserve. I have rarely seen the trees and tall herbs, such as nettles and teasels, grow so luxuriantly. This suggests that the vegetation in Swavesey is normally under stress due to shortage of water. Seedling trees in the wood (see pp. 43–45) benefited particularly from the wet winter. By the end of the summer of 2001 they had put on height and looked very healthy. The area of the wood's ground flora which is dominated by ivy (see p. 42) has increased considerably. This may be a direct effect of increased rainfall, or an indirect one resulting from cow parsley being shaded out by the more luxuriant tree foliage, or both. Our one woodland butterfly, the speckled wood (see pp. 64, 69), became noticeably commoner in the summer of 2001. Until that year, I had never counted more than four butterflies on one transect, but in 2001 I counted more than four butterflies on fifteen occasions. The largest number on one transect was eight. In 2001 speckled woods appeared to have four broods between 4th May and 4th October.

There was an interesting change in the newt populations in 2001. Smooth newts colonized the Mere in 1985 (see p. 61), two years after it was constructed. They remained common until 2000 when they became much rarer (see p. 62) and in 2001 none were observed. It seemed as if they had been eliminated by the grass snake, or snakes, which first appeared in 1998 (see p. 60). Despite the grass snakes, great crested newts appeared in 1999 (see p. 61). However, none were seen in 2000, but in 2001 they reappeared. It seems that grass snakes, possibly aided by great crested newts, have exterminated or greatly reduced the smooth newt population. Great crested newts are poisonous. This may be the reason why they have survived the presence of grass snakes.

Although reed warblers have visited the Mere in recent years (see p. 58), in 2001 the species was present continuously from 28th May to 3rd September. Even when it was not visible, I heard the male singing throughout this time. It probably bred but, until I cut the reeds next winter and find the nest, I cannot be certain. The 25th September 2001 was a red-letter day: when making my routine dragonfly transect I put up a jack snipe. It slipped away quietly over the southern bank of the Mere and did not return. Jack snipe are scarce visitors to Cambridgeshire and I had not seen one for many years. On 16th December 2001 our grand-daughter Esther found a dead harvest mouse in the garden – only our second record of this species.

I must now turn from the extremely parochial to the global scene. The terrible events of 11th September 2001 and the sequel force us to reassess everything, not least our approach to the conservation of people and their environment. There are likely to be immediate effects, both positive and negative. Warfare kills wildlife as well as people, although, as in Korea, no-man's-land can become a haven for wildlife. The reduction in air travel will for a time reduce the pollution of the atmosphere. The most important effect of the present crisis will be to take everyone's mind off matters not directly connected with it. This need not be a long-lasting effect. The outline of our national conservation policy, which resulted in our National Nature Reserves, National Parks and statutory conservation bodies, was put together in the darkest days of the Second World War.

Before 11th September the environmental scene was somewhat depressing. Rainforests were still being destroyed throughout the tropics; primaeval pine forests and the Arctic National Wildlife Refuge in Alaska were threatened by the new United States administration. Over-fishing, encouraged by the European Union, threatened to destroy the rich fishing grounds off Mauretania in western Africa, despite the establishment of the Banc d'Arguin National Park on the coast.

In Britain, foot-and-mouth disease threatened hill farming on which so much upland wildlife depends. The Government appeared to underestimate the potential hazards posed by growing genetically modified crops. The Prime Minister, Tony Blair, had promised to put the environment 'at the heart of Government', but this did not appear to be happening. However, although progress with British conservation legislation was slow, the conservation of habitats and species in the United Kingdom was increasingly strengthened by European legislation. Railways were getting worse and were not providing an adequate substitute for road transport. However, it was encouraging that the Minister in charge of transport, Mr Prescott, defended the fuel levy on the grounds that its long-term advantages far outweighed its present disadvantages.

The latest report of the Intergovernmental Committee on Climate Change made it clear that present climate changes were to a large extent attributable to man-made pollutants. Increasingly the measures proposed at Kyoto in 1997 were seen to be necessary. Therefore President George W. Bush's rejection of the Kyoto Treaty in March 2001 came as an affront to science and common sense. It was one of the most dispiriting international events of recent times. It very nearly destroyed the Kyoto Accord but, thanks to the far-sightedness of Mr Prescott and others, something was rescued from the wreckage and at least a framework exists for effective international and national action in the future.

Since 11th September, terrorism is seen to be a global problem which can be defeated only by global measures. It is greatly to Mr Blair's credit that, in the speech which he gave just after the disaster, he made it clear that the global co-operation now operating against terrorism should be extended to cover measures to combat climate change. Thus something immensely positive and important for all mankind could come out of the appalling disaster of 11th September. Of course, a renewed and strengthened determination to control pollutants causing climate change will not last unless there is strong political will and public pressure behind it. We all have a huge responsibility to support the new opportunities. Never has informed public debate about the environment been more important. I suggest that one of the best ways to get it is to ensure that the concept of Future Care becomes fully established in the political life of all nations.

# APPENDIX ONE

## Practical points for establishing and managing a small nature reserve

### Introduction

I hope that readers of this book will be encouraged to turn parts of their gardens into nature reserves, if they have not done so already. Of course, the extent that this can be achieved does depend on the amount of land which is available. I have been most fortunate in having one and a half acres to spare: similar results to mine can only be expected in large gardens or on farms. Nevertheless much can be achieved in small gardens. Ponds are particularly good at providing a home for a wealth of species in a small space. The number of birds which a small garden can harbour on its own is very limited, but birds do not recognize property boundaries and, if several contiguous gardens all provide some bushes and trees, the total area can support a good number of different species. Butterflies will feed on a wide range of garden flowers and shrubs, but their caterpillars depend upon the provision of wild plants, notably wild grasses such as meadow grass, fescues and couch; crucifers such as cuckoo-flower and garlic mustard; and nettles. These and other wild plants have to be provided if a garden is to support breeding populations of butterflies. To the dedicated gardener wild plants may look untidy, but it is the untidy bits of gardens and farms which support most wildlife.

Wild plants and animals will colonize any bit of land which is taken out of crop production. If the land is then planted with trees and bushes, or a pond is dug in it, many more species will turn up. Some gardeners and farmers will then be happy to see what comes and goes as the habitats develop, but many will have special species in mind, which they hope will colonize their reserve. They will then wish to manage their land for these species. However it will soon become obvious that having some species prevents them having others. Take my reserve as an example: reeds and blackthorns are particularly delightful plants, but if I do not control reeds on the Mere I will lose my dragonflies, and if I do not control blackthorn suckers I will lose most of the butterflies in the grassland of the Mere Enclosure. If I had not set and followed my management objectives the natural development of both the Mere and the meadow would have already caused the loss of the species which I most wanted to retain.

### Keeping records

To discover whether or not objectives are being achieved, tabs have to be kept on what happens year by year. Memory is very fallible: if no records are kept either of management or of how special species have fared it is difficult to know when to act and what to do. Therefore keeping notes about the state and maintenance of the reserve and about its species is a crucial part of reserve management. This may seem very obvious, but it is not; even on many well-known nature reserves the effects of management are inadequately monitored. Keeping simple lists of the species which are observed each year is essential. Making censuses and transects is necessary to discover more subtle changes. Quite apart from providing crucial facts for management, keeping notes adds enormously to the pleasure of having a nature reserve. It is fascinating and rewarding to record progress over the years as one enjoys the arrival of new species and the establishment of others.

## Scope and relevance of the advice given

The general observations made above provide the background for summarizing the particular lessons which I have learnt while establishing and managing my own reserve in Swavesey. I should emphasise that there are things which I have not yet done but which will have to be done in the future. For example, I shall have to coppice some of the trees to let more light and warmth into parts of the Wood. Eventually it will be necessary to dig out some of the sediment which accumulates in the Mere. Excellent practical advice about the details of management of land for wildlife can be obtained from numerous books, journals and leaflets. A small selection is given in Reference books, see page 123.

No two places are exactly alike; what works in one will not work in another. Differences in latitude, altitude, topography, rainfall and soils can make all the difference. What follows is a summary of the lessons which I have learnt in making my small nature reserve on heavy clay in low-lying flat country in East Anglia, where the rainfall is exceptionally low. I hope that, by summarizing my experience and making suggestions, these notes will help others, especially those establishing wildlife habitats in similar areas.

## Hardwood Plantation

### (i) *Tree species*

Of the 43 species planted, hybrid black-poplars and the hybrid conifer *Cupressocyparis leylandii* have grown the fastest. Of the native species planted, those commonly found in ancient woods in the vicinity have done best, i.e. oak, ash, field maple, hazel, willow and hawthorn. Hornbeam and small-leaved lime have also thrived. Drought has had severe effects on birch, beech and service, and on some willows and ashes. Leylandii and holm oak have suffered badly from exceptionally severe gales. Birds which require evergreens, especially in winter, can be catered for by allowing ivy to grow thickly on some trees. Young woods do not provide nesting sites for hole-nesting birds; these can only be provided by installing nest boxes.

### (ii) *Planting trees on grassland*

Trees planted when small (*c.* two feet high) have done better than those planted when they are older (*c.* six feet high). The small ones often overtake the older ones in a few years. This is presumably because the root/stem ratio is nearly always much better in small trees. Small trees are much cheaper to buy, especially if they are obtained from wholesalers and nurserymen and not from garden centres.

The larger the area dug before planting the less root competition does the young tree get from grasses. Subsequently, grass can be discouraged by cutting it and covering the area round the trees with the cut grass or a mulch. In dry seasons recently planted young trees require watering, at least during the first season. Planting trees about six feet apart soon causes the grass to be shaded out and encourages the trees to grow taller and faster as they compete for light. Later, as the trees grow older, competition for light and water causes some to die. This natural thinning produces dead wood, which is valuable for many species of invertebrates. When one wishes to protect a particular individual tree it is sometimes necessary to fell or cut back its neighbour.

The only prominent disorder which I have observed, apart from Dutch Elm Disease, is a gall on acorns. I did nothing about it and it soon disappeared. It returned some years later and then disappeared again.

### (iii) Woodland ground flora

Cow parsley, ivy and cleavers quickly dominate the ground flora once the original grass species have been shaded out. Characteristic species of ancient woodland such as bluebells, dog's mercury, primroses and wood anemone will grow and even spread beneath the planted trees, but only if cow parsley, ivy and cleavers are laboriously weeded out, i.e. by 'gardening'.

### Grassland

If grassland is left ungrazed by stock it lasts for a surprisingly long time but develops into tussocks which are difficult to mow. Areas of grass which are to be kept short should, therefore, be mowed regularly from the time when the land ceases to be grazed. Woody plants – mostly hawthorn, wild roses and elder – do colonize grassland, but the main threat is from blackthorn suckers, which become very numerous once blackthorn bushes in the vicinity become well established.

Roadside verges in the district provide a useful indication of what ground flora can be encouraged to grow in a reserve's ungrazed grassland. At Swavesey, oxeye daisy, meadow crane's-bill and teasel have been successfully introduced and have spread. Thistles (along with brambles) are valuable food plants for adult butterflies. If they increase too much they must be controlled by pulling when the soil is damp enough.

### Pond

Heavy clay makes digging very hard work and it can both rot and dessicate plants according to its moisture content. However, it provides wonderful opportunities for creating ponds. This is because a pond dug in impermeable clay does not require lining, which is very expensive. If you are on clay you can afford to make a much larger pond than would otherwise be the case.

Many ponds which have been constructed by damming a stream become polluted by fertilizer run-off and develop algal blooms. A pond which has no input from polluted water is much less likely to have that problem. Nevertheless blooms may still occur although in my experience they are very transient. I suspect that this has been due to the rich invertebrate fauna of the Mere, which includes several species of water snails, which I introduced in 1984. Unlike dragonflies, and most other aquatic invertebrates, water snails are poor dispersers and so need help in colonizing a new pond. I have never had to put bales of barley straw into the Mere to control algae, but there is growing evidence to show that this is an effective measure when required.

Aquatic plants which are found locally have done well on the Mere when introduced. The plants which have colonized it naturally – reed and bulrush – have caused management problems. So long as it retained its normal water-level a large part of it was too deep for these species (i.e. four to five feet) but, with prolonged drought and the subsequent drop in the water-level, reeds – and to a lesser extent bulrushes – were able to colonize most of the pond, and I can keep them at bay only by laboriously cutting them.

An unwelcome colonist has been the alien New Zealand pygmyweed or water stonecrop (*Crassula helmsii*). It is very abundant on the edges of the Mere. Several species, notably Canadian pondweed, floating pondweed and common water-starwort, have declined from peaks of abundance. It is possible that this has been due to the New Zealand pygmyweed. Therefore, as a safety measure, it would probably be best to eradicate this species as soon as it appears.

Ponds provide interesting habitats at all stages of their development but if, like me, you want your pond to support dragonflies, a host of other aquatic invertebrates and newts, one must be prepared to carry out a considerable amount of management. This is because keeping a pond at early stages of development is working against nature. Scrub on the edges of the pond must be cut to prevent too much shading and also to prevent the pond collecting too many dead leaves. This work can be minimized by not planting too many trees and shrubs by the water's edge when the pond is created. Aggressive water-weeds like reeds and bulrushes will need annual cutting to allow other water plants to survive and to keep the edges of the pond warm. Fish, particularly if present in large numbers, can greatly reduce the numbers and richness of the invertebrate fauna: hard choices have to be made.

### Fencing

Fencing is necessary to keep stock out until hedges have become thick enough to be stock-proof. When laying out the fence line, the length of a cow's or horse's neck should not be underestimated! Wooden stakes four feet high with four strands of wire (the three top ones barbed) placed four feet from the hedge plants have proved adequate at Swavesey. Even after 16 years, the hedges round the Mere would probably not be completely stock-proof without the support of the original fencing. The hedges are effectively trimmed by my neighbour's cattle. They are now about 12 feet high and will soon need cutting so that their bottoms remain thick and suitable for nesting birds.

# APPENDIX TWO

## Birds observed from our land at Swavesey 1960–2001

Species which have bred are in **bold:**

Grey Heron
Spoonbill
Common Teal
Mute Swan
**Mallard**
Sparrowhawk
Kestrel
Hobby
**Red-legged Partridge**
Grey Partridge
Pheasant
**Moorhen**
Golden Plover
Lapwing
Jack Snipe
Common Snipe
Woodcock
Black-headed Gull
Common Gull
Lesser Black-backed Gull
Herring Gull
Great Black-backed Gull
**Stock Dove**
**Wood Pigeon**
**Collared Dove**
**Turtle Dove**
Cuckoo
**Little Owl**

**Tawny Owl**
Common Swift
Kingfisher
Green Woodpecker
Great Spotted Woodpecker
Skylark
Sand Martin
**Swallow**
House Martin
Meadow Pipit
Yellow Wagtail
Grey Wagtail
Pied Wagtail
**Wren**
**Dunnock**
**Robin**
Whinchat
**Blackbird**
Fieldfare
**Song Thrush**
Mistle Thrush
Redwing
Sedge Warbler
Reed Warbler
**Lesser Whitethroat**
**Whitethroat**
**Blackcap**
Chiffchaff

**Willow Warbler**
Goldcrest
**Spotted Flycatcher**
Pied Flycatcher
**Long-tailed Tit**
Coal Tit
**Blue Tit**
**Great Tit**
Treecreeper
Jay
**Magpie**
**Jackdaw**
**Rook**
Carrion Crow
**Starling**
**House Sparrow**
**Tree Sparrow**
**Chaffinch**
**Greenfinch**
**Goldfinch**
**Linnet**
**Bullfinch**
Crossbill
Yellowhammer
Corn Bunting
Reed Bunting

# APPENDIX THREE

## Scientific names of species mentioned in the text

### Notes

(i) The species listed, except for those in square brackets, either occurred on my land before the reserve was established; or are plants introduced by me; or are plants and animals which have colonized the reserve unaided.

(ii) * An asterisk indicates species native to Britain but probably not native to clay areas of Cambridgeshire.

(iii) [ ] Square brackets indicate other species mentioned in the text but not occurring in my reserve.

### 1. Species native to Britain or naturalized in Britain

### Plants

*Chara* sp. (Stonewort)
*Pinus sylvestris* L. (Scots Pine)
*Taxus baccata* L. (Yew)
*Nymphaea alba* L. (White Water-lily)
*Caltha palustris* L. (Marsh-marigold, Kingcup)
*Anemone nemorosa* L. (Wood Anemone)
*Ranunculus lingua* L. (Greater Spearwort)
*Ranunculus ficaria* L. (Lesser Celandine)
*Ranunculus peltatus* Schrank (Pond Water-crowfoot)
*Ulmus procera* Salisb. (English Elm)
*Humulus lupulus* L. (Hop)
*Urtica dioica* L. (Common Nettle)
*Fagus sylvatica* L. (Beech)
*Castanea sativa* Mill. (Sweet Chestnut)

*\*Quercus petraea* (Matt.) Liebl. (Sessile Oak)
*Quercus robur* L. (Pedunculate Oak)
*Betula pendula* Roth (Silver Birch)
*Alnus glutinosa* (L.) Gaertn. (Alder)
*Carpinus betulus* L. (Hornbeam)
*Corylus avellana* L. (Hazel)
*Tilia cordata* Mill. (Small-leaved Lime)
*Viola odorata* L. (Sweet Violet)
*Populus alba* L. (White Poplar)
*Populus tremula* L. (Aspen)
*Populus nigra* L. (Black-poplar)
*Salix fragilis* L. (Crack-willow)
*Alliaria petiolata* (M. Bieb) Cavara & Grande (Garlic Mustard)
*Cardamine pratensis* L. (Cuckooflower, Lady-smock)
*Primula vulgaris* Huds. (Primrose)
*Primula elatior* (L.) (Oxlip)
*Primula veris* L. (Cowslip)
*Lysimachia nummularia* L. (Creeping-Jenny)
*Crassula helmsii* (Kirk) Cockayne (New Zealand Pigmyweed, New Zealand Water Stonecrop)
*Rubus* spp. (Brambles)
*Rosa canina* L. (Dog-rose)
*Prunus cerasifera* Ehrh. (Cherry Plum)
*Prunus spinosa* L. (Blackthorn)
*Prunus avium* (L.) (Gean, Wild Cherry)
*Prunus padus* L. (Bird Cherry)
*\*Sorbus aucuparia* L. (Rowan)
*Sorbus torminalis* (L.) (Wild Service-tree)
*Crataegus monogyna* Jacq. (Hawthorn)
*Lotus corniculatus* L. (Common Bird's-foot-trefoil)
*\*Ilex aquifolium* L. (Holly)

– 114 –

*Mercurialis perennis* L. (Dog's Mercury)
*Rhamnus cathartica* L. (Buckthorn)
*Frangula alnus* Mill. (Alder Buckthorn)
*Acer campestre* L. (Field Maple)
*Acer pseudoplatanus* L. (Sycamore)
*Geranium* sp.
*Geranium pratense* L. (Meadow
   Crane's-bill)
*Hedera helix* L. (Ivy)
*Anthriscus sylvestris* (L.) Hoffm.
   (Cow Parsley)
*Conium maculatum* L. (Hemlock)
*Heracleum sphondylium* L. (Hogweed)
* *Vinca minor* L. (Lesser Periwinkle)
* *Menyanthes trifoliata* L. (Bogbean)
*Stachys sylvatica* L. (Wood or Hedge
   Woundwort)
*Glechoma hederacea* L. (Ground-ivy)
*Lycopus europaeus* L. (Gypsywort)
*Hippuris vulgaris* L. (Mare's-tail)
*Callitriche stagnalis* Scop. (Common
   Water-starwort)
*Fraxinus excelsior* L. (Ash)
*Ligustrum vulgare* L. (Privet)
*Galium aparine* L. (Cleavers,
   Goose-grass)
*Sambucus nigra* L. (Elder)
*Lonicera periclymenum* L.
   (Honeysuckle)
*Dipsacus fullonum* L. (Wild Teasel)
*Cirsium vulgare* (Savi) Ten. (Spear
   Thistle)
*Cirsium arvense* (L.) Scop. (Creeping
   Thistle)
*Picris echioides* L. (Bristly Oxtongue)
*Crepis biennis* L. (Rough or Greater
   Hawk's-beard)
*Artemisia vulgaris* L. (Mugwort)
*Achillea millefolium* L. (Yarrow)
*Tussilago farfara* L. (Colt's-foot)
*Butomus umbellatus* L. (Flowering-rush)
*Alisma plantago-aquatica* L. (Water-
   plantain)
* *Hydrocharis morsus-ranae* L. (Frogbit)
* *Stratiotes aloides* L. (Water-soldier)

*Elodea canadensis* Michx. (Canadian
   Waterweed)
*Potamogeton natans* L. (Broad-leaved or
   Floating Pondweed)
*Potamogeton crispus* L. (Curled
   Pondweed)
*Arum maculatum* L. (Lords-and-ladies)
*Juncus effusus* L. (Soft-rush)
*Carex* spp. (Sedges)
*Festuca rubra* L. (Red Fescue)
*Lolium perenne* L. (Perennial Rye-grass)
*Dactylis glomerata* L. (Cock's-foot)
*Arrhenatherum elatius* (L.) P. Beauv.
   ex J. Presl. & C. Presl. (False Oat-
   grass)
*Holcus lanatus* L. (Yorkshire-fog)
*Agrostis stolonifera* L. (Creeping Bent)
*Alopecurus pratensis* L. (Meadow
   Foxtail)
*Elytrigia repens* (L.) Desv. ex Nevski
   (Common Couch)
*Hordeum secalinum* Schreb. (Meadow
   Barley)
*Phragmites australis* (Cav.) Trin. ex
   Steud. (Common Reed)
*Sparganium erectum* L. (Branched
   Bur-reed)
*Typha latifolia* L. (Bulrush, Great
   Reedmace)
* *Leucojum aestivum* L. (Summer
Snowflake)
*Iris pseudacorus* L. (Yellow Iris, Yellow
   Flag)
*Iris foetidissima* L. (Stinking Iris)
*Narcissus pseudonarcissus* L.
   (Wild Daffodil)

**Invertebrates**

MOLLUSCA
*Lymnaea stagnalis* (L.) (Great Pond-
   snail)
[*Mytilus edulis* L. (Edible Mussel)]
[*Ostrea discoides* (a fossil oyster)]

ODONATA
[*Calopteryx virgo* (L.) (Beautiful
Demoiselle)]
*Calopteryx splendens* (Harris) (Banded
Demoiselle)
*Lestes sponsa* (Hansem.) (Emerald
Damselfly)
[*Platycnemis pennipes* (Pallas) (White-
legged Damselfly)]
*Pyrrhosoma nymphula* (Sulz.) (Large
Red Damselfly)
*Erythromma najas* (Hansem.) (Red-eyed
Damselfly)
*Coenagrion puella* (L.) (Azure
Damselfly)
*Coenagrion pulchellum* (Vander Lind.)
(Variable Damselfly)
*Enallagma cyathigerum* (Charp.)
(Common Blue Damselfly)
*Ischnura elegans* (Vander Lind.)
(Blue-tailed Damselfly)
*Aeshna mixta* Latr. (Migrant Hawker)
*Aeshna cyanea* (O. F. Müll.) (Southern
Hawker)
*Aeshna grandis* (L.) (Brown Hawker)
*Anax imperator* Leach (Emperor
Dragonfly)
*Brachytron pratense* (O. F. Müll.)
(Hairy Dragonfly)
*Libellula quadrimaculata* L.
(Four-spotted Chaser)
*Libellula depressa* L. (Broad-bodied
Chaser)
*Libellula fulva* (O. F. Müll.) (Scarce
Chaser)
*Orthetrum cancellatum* (L.) (Black-
tailed Skimmer)
*Sympetrum striolatum* (Charp.)
(Common Darter)
*Sympetrum sanguineum* (O. F. Müll.)
(Ruddy Darter)

ORTHOPTERA
*Meconema thalassinum* (De Geer)
(Oak Bush-cricket)

*Chorthippus albomarginatus* (De Geer)
(Lesser Marsh Grasshopper)

HEMIPTERA
*Gerris* sp. (Pond-skater)
*Nepa cinerea* L. (Water-scorpion)
*Ranatra linearis* (L.) (Water Stick-insect)
*Notonecta* sp. (Back-swimmer, Water-
boatman)
*Corixa* spp. (Water-boatmen, Lesser
Water-boatmen)

RAPHIDIOPTERA
*Xanthostigma* sp. (Snake Fly)

MEGALOPTERA
*Sialis lutaria* (L.) (Alder Fly)

MECOPTERA
*Panorpa communis* L. (Scorpion Fly)

LEPIDOPTERA
*Thymelicus sylvestris* (Poda) (Small
Skipper)
*Thymelicus lineola* (Ochsen.) (Essex
Skipper)
*Ochlodes faunus* (Turati) (Large
Skipper)
[*Papilio machaon* (L.) (Swallowtail)]
*Gonepteryx rhamni* (L.) (Brimstone)
*Pieris brassicae* (L.) (Large White)
*Pieris rapae* (L.) (Small White)
*Pieris napi* (L.) (Green-veined White)
[*Pontia daplidice* (L.) (Bath White)]
*Anthocharis cardamines* (L.)
(Orange-tip)
[*Callophrys rubi* (L.) (Green
Hairstreak)]
*Satyrium w-album* (Knoch) (White-letter
Hairstreak)
[*Thecla betulae* (L.) (Brown Hairstreak)
[*Quercusia quercus* (L.) (Purple
Hairstreak)]
[*Strymonidia pruni* (L.) (Black
Hairstreak)]
*Lycaena phlaeas* (L.) (Small Copper)

*Aricia agestis* (D.&S.) (Brown Argus)
*Polyommatus icarus* (Rott.) (Common Blue)
*Celastrina argiolus* (L.) (Holly Blue)
[*Limenitis camilla* (L.) (White Admiral)]
*Vanessa atalanta* (L.) (Red Admiral)
*Aglais urticae* (L.) (Small Tortoiseshell)
*Inachis io* (L.) (Peacock)
*Polygonia c-album* (L.) (Comma)
*Pararge aegeria* (L.) (Speckled Wood)
*Lasiommata megera* (L.) (Wall)
*Pyronia tithonus* (L.) (Gatekeeper)
*Maniola jurtina* (L.) (Meadow Brown)
*Coenonympha pamphilus* (L.) (Small Heath)
*Aphantopus hyperantus* (L.) (Ringlet)
*Mimas tiliae* (L.) (Lime Hawk-moth)
*Smerinthus ocellata* (L.) (Eyed Hawk-moth)
*Laothoe populi* (L.) (Poplar Hawk-moth)
*Deilephila elpenor* (L.) (Elephant Hawk-moth)
*Cerura vinula* L. (Puss Moth)
*Calliteara pudibunda* (L.) (Pale Tussock)
*Autographa gamma* (L.) (Silver Y)
*Catocala nupta* (L.) (Red Underwing)

COLEOPTERA
*Dytiscus marginalis* L. (Great Water-beetle)
*Gyrinus* sp. (Whirligig Beetle)
*Scolytus* sp. (Bark Beetle)

HYMENOPTERA
*Vespa crabro* L. (Hornet)
*Bombus* spp. (Bumble Bees)

**Vertebrates**

FISHES
*Gasterosteus aculeatus* L. (Three-spined Stickleback)

AMPHIBIANS
*Triturus vulgaris* (L.) (Smooth Newt)

*Triturus cristatus* (Laurenti) (Great Crested Newt)
*Bufo bufo* (L.) (Common Toad)
*Rana temporaria* L. (Common Frog)

REPTILES
*Natrix natrix* (L.) (Grass Snake)

BIRDS
[*Tachybaptus ruficollis* (Pall.) (Little Grebe)]
[*Podiceps cristatus* (L.) (Great Crested Grebe)]
*Ardea cinerea* L. (Grey Heron)
[*Nycticorax nycticorax* (L.) (Black-crowned Night Heron)]
*Platalea leucorodia* L. (Spoonbill)
*Cygnus olor* (Gmel.) (Mute Swan)
[*Cygnus columbianus bewickii* (Yarr.) (Bewick's Swan)]
[*Tadorna tadorna* (L.) (Shelduck)]
[*Anas penelope* L. (Wigeon)]
*Anas crecca* L. (Common Teal)
*Anas platyrhynchos* L. (Mallard)
[*Anas acuta* L. (Pintail)]
[*Aythya fuligula* (L.) (Tufted Duck)]
*Accipiter nisus* (L.) (Sparrowhawk)
*Falco tinnunculus* L. (Kestrel)
*Falco subbuteo* L. (Hobby)
*Alectoris rufa* (L.) (Red-legged Partridge)
*Perdix perdix* (L.) (Grey Partridge)
*Phasianus colchicus* L. (Pheasant)
*Gallinula chloropus* (L.) (Moorhen)
*Fulica atra* L. (Coot)
[*Haematopus ostralegus* L. (Oystercatcher)]
[*Charadrius hiaticula* L. (Ringed Plover)]
*Pluvialis apricaria* (L.) (Golden Plover)
*Vanellus vanellus* (L.) (Lapwing)
[*Philomachus pugnax* (L.) (Ruff)]
*Lymnocryptes minimus* (Brünn.) (Jack Snipe)

*Gallinago gallinago* (L.) (Common Snipe)

*Scolopax rusticola* L. (Woodcock)

[*Limosa limosa* (L.) (Black-tailed Godwit)]

[*Tringa totanus* (L.) (Redshank)]

*Larus ridibundus* L. (Black-headed Gull)

*Larus canus* L. (Common Gull)

*Larus fuscus* L. (Lesser Black-backed Gull)

*Larus argentatus* Pontopp. (Herring Gull)

*Larus marinus* L. (Great Black-backed Gull)

[*Sterna hirundo* L. (Common Tern)]

*Columba oenas* L. (Stock Dove)

*Columba palumbus* L. (Wood Pigeon)

*Streptopelia decaocto* (Frivalds.) (Collared Dove)

*Streptopelia turtur* (L.) (Turtle Dove)

*Cuculus canorus* L. (Cuckoo)

*Athene noctua* (Scop.) (Little Owl)

*Strix aluco* L. (Tawny Owl)

*Apus apus* (L.) (Common Swift)

*Alcedo atthis* (L.) (Kingfisher)

*Picus viridis* L (Green Woodpecker)

*Dendrocopos major* (L.) (Great Spotted Woodpecker)

*Alauda arvensis* L. (Skylark)

*Riparia riparia* (L.) (Sand Martin)

*Hirundo rustica* L. (Swallow)

*Delichon urbica* (L.) (House Martin)

*Anthus pratensis* (L.) (Meadow Pipit)

*Motacilla flava flavissima* (Blyth) (Yellow Wagtail)

*Motacilla cinerea* Tunst. (Grey Wagtail)

*Motacilla alba yarrellii* Gould (Pied Wagtail)

*Troglodytes troglodytes* (L.) (Wren)

*Prunella modularis* (L.) (Dunnock)

*Erithacus rubecula* (L.) (Robin)

*Saxicola rubetra* (L.) (Whinchat)

*Turdus merula* L. (Blackbird)

*Turdus pilaris* L. (Fieldfare)

*Turdus philomelos* C.L. Brehm (Song Thrush)

*Turdus iliacus* L. (Redwing)

*Turdus viscivorus* L. (Mistle Thrush)

*Acrocephalus schoenobaenus* (L.) (Sedge Warbler)

*Acrocephalus scirpaceus* (Hermann) (Reed Warbler)

*Sylvia curruca* (L.) (Lesser Whitethroat)

*Sylvia communis* Latham (Common Whitethroat))

*Sylvia atricapilla* (L.) (Blackcap)

*Phylloscopus collybita* (Vieill.) (Chiffchaff)

*Phylloscopus trochilus* (L.) (Willow Warbler)

*Regulus regulus* (L.) (Goldcrest)

*Muscicapa striata* (Pall.) (Spotted Flycatcher)

*Ficedula hypoleuca* (Pall.) (Pied Flycatcher)

*Aegithalos caudatus* (L.) (Long-tailed Tit)

*Parus ater* L. (Coal Tit)

*Parus caeruleus* L. (Blue Tit)

*Parus major* L. (Great Tit)

[*Sitta europaea* L. (Nuthatch)]

*Certhia familiaris* L. (Treecreeper)

*Garrulus glandarius* (L.) (Jay)

*Pica pica* (L.) (Magpie)

*Corvus monedula* L. (Jackdaw)

*Corvus frugilegus* L. (Rook)

*Corvus corone corone* L. (Carrion Crow)

*Sturnus vulgaris* L. (Starling)

*Passer domesticus* (L.) (House Sparrow)

*Passer montanus* (L.) (Tree Sparrow)

*Fringilla coelebs* L. (Chaffinch)

*Carduelis chloris* (L.) (Greenfinch)

*Carduelis carduelis* (L.) (Goldfinch)

*Carduelis cannabina* (L.) (Linnet)

*Loxia curvirostra* L. (Common Crossbill)

*Pyrrhula pyrrhula* (L.) (Bullfinch)

*Emberiza citrinella* L. (Yellowhammer)
*Emberiza schoeniclus* (L.) (Reed Bunting)
*Miliaria calandra* (L.) (Corn Bunting)

MAMMALS

*Plecotus auritus* (L.) (Brown Long-eared Bat)
*Pipistrellus pipistrellus* Kaup (Pipistrelle)
*Erinaceus europaeus* L. (Hedgehog)
*Talpa europaea* L. (Mole)
*Sorex araneus* L. (Common Shrew)
*Sorex minutus* L. (Pigmy Shrew)
*Neomys fodiens* Schreb. (Water Shrew)
*Vulpes vulpes* (L.) (Fox)
*Mustela erminea* L. (Stoat)
*Meles meles* (L.) (Badger)
[*Phoca vitulina* L. (Common Seal)]
*Lepus europaeus* Pall. (Brown Hare)
*Oryctolagus cuniculus* (L.) (Rabbit)
*Sciurus carolinensis* Gmel. (Grey Squirrel)
*Micromys minutus* (Pall.) (Harvest Mouse)
*Apodemus sylvaticus* (L.) (Wood Mouse)
*Mus musculus* L. (House Mouse)
*Rattus norvegicus* (Erxl.) (Brown Rat)
*Clethrionomys glareolus* Schreb. (Bank Vole)
*Arvicola terrestris* (L.) (Water Vole)
*Microtus agrestis* (L.) (Field or Short-tailed Vole)

## 2. Non-native species

### Plants

*Picea abies* (L.) H. Karst. (Norway Spruce)
*Larix decidua* Mill. (European Larch)
*Metasequoia glyptostroboides* Hu & W.C. Cheng (Dawn Redwood)
×*Cupressocyparis leylandii* (A.B. Jacks. & Dallim.) (Leyland Cypress)

*Ginkgo biloba* L. (Maidenhair Tree)
*Juglans regia* L. (Walnut)
*Quercus ilex* L. (Evergreen Oak, Holm Oak)
*Alnus rubra* Bong. (Red Alder)
*Populus nigra* var. *italica* Muenchh. (Lombardy Poplar)
*Populus* × *canadensis* Moench (Hybrid Black-poplar)
*Pyracantha coccinea* M. Roem. (Firethorn)
*Aesculus indica* (Cambess.) (Indian Horse-chestnut)
*Aesculus hippocastanum* L. (Horse-chestnut)
*Vinca major* L. (Greater Periwinkle)
[*Astelia* sp.]

### Animals

[*Megalagrion* sp. (a genus of endemic Hawaiian damselflies)]
[*Hypopetalia pestilens* McLachl. (a Chilean neopetaliid dragonfly)]
[*Phenes raptor* Rambur (a Chilean petalurid dragonfly)]
[*Melopsittacus undulatus* (Shaw & Nodder) (Budgerigar)]
[*Thylacinus cynocephalus* (Harris) (Tasmanian Wolf, Thylacine)]

N.B. Domestic cultivars of apples and pears are not included in the list.

Scientific and English names in the list of flora are based on Stace, C. A. (1997). *New flora of the British Isles* (edn 2). Cambridge, University Press. Those in the lists of fauna are based on Friday, L. E. & Harley, B. H. (2000). *Checklist of the flora and fauna of Wicken Fen.* Colchester, Harley Books.

# REFERENCES

## (Literature referred to in the text or relevant to it)

**Preface**

Moore, N. W. (1987). *The Bird of Time – the science and politics of nature conservation.* Cambridge, University Press.

**Introduction**

Moore, N. W. (compiler) (1997). *Dragonflies – status survey and conservation action plan.* Gland, Switzerland and Cambridge, IUCN.

**Chapter 1**

Moore, N. W. (1962). The heaths of Dorset and their conservation. *Journal of Ecology* 50: 369–391.

Sheail, J. (1985). *Pesticides and nature conservation. The British experience 1950–1975.* Oxford, Clarendon Press.

**Chapter 2**

Bircham, P. M. M. (1989). *The birds of Cambridgeshire.* Cambridge, University Press.

Ravensdale, J. R. (1982). *History on your doorstep.* London, British Broadcasting Corporation.

**Chapter 3**

*Enclosure Award Map. Swavesey.* (1838).

Ordnance Survey (Reprint 1970). *Old series of Ordnance Survey one-inch maps. Cambridge.* Newton Abbot, David and Charles.

*Tythe map. Swavesey.* (1840).

**Chapter 6**

Moore, N. W. & Hooper, M. D. (1975). On the number of bird species in British woods. *Biological Conservation* 8: 239–250.

Perring, F. H., Sell, P. D., Walters, S. M. & Whitehouse, H. L. K. (1964). *A flora of Cambridgeshire.* Cambridge, University Press.

Peterken, G. F. & Harding, P. T. (1975). Woodland conservation in eastern England: comparing the effects of changes in three study areas since 1946. *Biological Conservation* 8: 279–298.

**Chapter 7**

Corbet, P. S., Longfield, C. & Moore, N. W. (1960). *Dragonflies.* London, Collins.

Longfield, C. (1937). *The dragonflies of the British Isles.* London, Frederick Warne.

**Chapter 8 (B)**

McArthur, R. H. & Wilson, E. O. (1967). *The theory of biogeography.* Princeton, University Press.

Moore, N. W. & Hooper, M. D. (1975). On the number of bird species in British woods. *Biological Conservation* 8: 239–250.

**Chapter 8 (D)**

Bennett, T. & Perrin, V. (1994). The butterflies of Cambridgeshire: high-lights of a county survey (1985–1992). *Nature in Cambridgeshire* 36: 3–17.

Moore, N. W. (1975). Butterfly transects in a linear habitat 1964–1973. *Entomologist's Gazette* **26**: 71–78.

Pollard, E. (1979). A national scheme for monitoring the abundance of butterflies: the first three years. *Proceedings and Transactions of the British Entomological and Natural History Society* **12**: 77–90.

—, Elias, D. O., Skelton, M. J. & Thomas, J. A. (1975). A method of assessing the abundance of butterflies in Monks Wood National Nature Reserve in 1973. *Entomologist's Gazette* **26**: 79–88.

Sparks, T. & Greatorex-Davies, N. (1997). The butterfly monitoring scheme in Cambridgeshire – coming of age. *Nature in Cambridgeshire* **39**: 22–32.

Thomas, J. A. (1984). The conservation of butterflies. *In* Vane Wright, R. I. & Ackery, P. R. (eds), *Biology of butterflies*, pp. 333–353. Proceedings of the Royal Entomological Society 11. London, Academic Press.

**Chapter 8 (E)**

Moore N. W. (1964). Intra- and interspecific competition among dragonflies (Odoments on population density control in Dorset, 1954–60. *Journal of Animal Ecology* **33**: 49–71.

— (1991). Male *Sympetrum striolatum* (Charp.) 'defends' a basking spot rather than a particular locality (Anisoptera: Libellulidae). *Notulae Odonatologicae* 3(7): 112.

— (1995) Experiments on population density of male *Coenagrion puella* (L.) by water (Zygoptera: Coenagrionidae). *Odonatologica* 24(1): 123–128.

— (2000). Interspecific encounters between male Aeshnids. Do they have a function? *Pantala International Journal of Odonatology* 3(2): 141–151.

Perrin, V. & Johnson, I. (1995). The Cambridgeshire Dragonfly Survey. *Nature in Cambridgeshire* **37**: 8–19.

**Chapter 8 (F)**

Friday, L. E. (ed.) (1997). *Wicken Fen – the making of a wetland nature reserve*. Colchester, Harley Books.

— & Harley, B. H. (2000). *Checklist of the flora and fauna of Wicken Fen*. Colchester, Harley Books.

**Chapter 9**

Moore, N. W. (1962). The heaths of Dorset and their conservation. *Journal of Ecology* **50**: 369–391.

Pollard, E., Hooper, M. D. & Moore, N. W. (1974). *Hedges*. London, Collins.

**Chapter 10**

Farming and Wildlife Advisory Group (FWAG). *Annual Reviews*.

Bentley-Smith, R. (2000). The Farming and Wildlife Advisory Group (FWAG) – an organisation for the 21st century. *Journal of the Royal Agricultural Society of England*. November 2000: 90–99.

O'Connor, R. J. & Shrub, M. (1986). *Farming and birds*. Cambridge, University Press.

*Reports of the Nature Conservancy* (1950–1972), *Nature Conservancy Council* (1973–1990) *and English Nature* (1991–).

## Chapter 11

Rackham, O. (1976). *Trees and woodland in the British landscape.* London, Dent.

## Chapter 12

Moore, N. W. (1991). Recent developments in the conservation of Odonata in Great Britain. *Advances in Odonatology* 5: 103–108.

## Chapter 14

Great Britain (1959). Final Act of the Conference on Antarctic together with the Antarctic Treaty. Miscellaneous No. 21 (1959) Command Paper 913, London, HMSO.

Great Britain (1961). The Antarctic Treaty (in four languages), presented by the Secretary of State for Foreign Affairs by Command of Her Majesty. Treaty Series No. 97 (1961). Command Paper 1535, London, HMSO.

## Chapter 15

IPCC (2001). Climate Change 2001: The Scientific Basis – Contribution of Working Group I to the Third Assessment Report of the Intergovernmental Panel on Climate Change. United Nations Environment Programme (UNEP) (through EarthPrint.com).

IPCC (2001). Climate Change 2001: Impacts, Adaptation and Vulnerability – Contribution of Working Group II to the Third Assessment Report of the Intergovernmental Panel on Climate Change. UNEP.

IPCC (2001). Climate Change 2001: Mitigation – Contribution of Working Group III to the Third Assessment Report of the Intergovernmental Panel on Climate Change. UNEP.

# Further Reading

## Identification

Barnard, P. C. (1999). *Identifying British insects and arachnids. An annnotated bibliography of key works.* Cambridge, University Press.

Brooks, S. (ed.), (1997). *Field guide to the dragonflies and damselflies of Great Britain and Ireland.* Hook, British Wildlife Publishing.

Clapham, A. R., Tutin, T. G. & Moore, D. M. (1987). *Flora of the British Isles* (edn 3). Cambridge, University Press.

Corbet, G. B. & Harris, S. (eds) (1991). *The handbook of British mammals* (edn 3). Oxford, Blackwell Scientific Publications.

Keble Martin, W. (1965). *The concise British flora in colour.* London, George Rainbird.

Lewington, R. (1999). *How to identify butterflies.* London, HarperCollins.

Miller, P. L. (1995). *Dragonflies* (edn 2). Naturalists' Handbooks 7. Slough, Richmond Publishing.

Peterson, R. T., Mountfort, G. & Hollom, P. A. D. (1993). *Birds of Britain and Europe* (edn 5). London, HarperCollins.

Philips, R. (1994). *Wildflowers of Britain.* Barnstaple & Oxford, Pan Books.

Thomas, J. A. (1986). RSNC Guide to *Butterflies of the British Isles.* Twickenham, Country Life Books.

Stace, C. A. (1997). *New flora of the British Isles* (edn. 2). Cambridge, University Press.

# REFERENCES

## Management for Wildlife

Baines, C. (1984). *How to make a wildlife garden*. London, Elm Tree Books.

Fry, R. & Lonsdale, D. (eds) (1991). *Habitat conservation for insects – a neglected green issue*. Middlesex, Amateur Entomologists' Society.

Johnson, H. & Johnson, P. (1999). *The birdwatcher's garden*. Lewes, Guild of Master Craftsmen Publishers.

Kirby, P. (1992). *Habitat management for invertebrates – a practical handbook*. Sandy, RSPB.

Sutherland, W. J. & Hill, D. A. (eds) (1995). *Managing habitats for conservation*. Cambridge, University Press.

**Note.** There are numerous manuals, guides and leaflets to help gardeners, farmers and reserve managers conserve habitats for wildlife, for example by The British Dragonfly Society, The British Trust for Conservation Volunteers, The Farming and Wildlife Advisory Group and The Royal Society for the Protection of Birds. The bi-monthly *British Wildlife* magazine frequently contains articles about conservation management.

*Animal* – a multicellular organism unable to manufacture its own food, therefore typically mobile, any organism which is not a plant, fungus or a micro-organism.

*Biodiversity* – the wide variety of species (with genetic variants) together with the habitats on which they depend.

*Boulder clay* – clay, originally incorporated in ice sheets, and later deposited on the land when the ice melted at the end of an Ice Age.

*Brown butterfly* – a butterfly belonging to the subfamily Satyrinae, for example meadow brown.

*Damselfly* – member of the suborder Zygoptera. Most damselflies are smaller than most Anisoptera or 'true dragonflies' (see below) and typically they perch with their wings folded above the back.

*Dragonfly* – member of the order Odonata. Dragonflies include the smaller damselflies (see above) and the larger Anisoptera or 'true dragonflies', whose wings are spread open when they perch.

*Drove* – a green lane or road used originally for driving cattle from summer grazing on fen pastures to high ground (and markets).

*Ecosystem* – a community of plants and animals and their physical environment, whose living and non-living parts interact to form a stable system.

*Fen* – vegetation growing on low-lying, water-logged, organic soils, which consist of alkaline or nearly neutral peat. The large intensively farmed area of eastern England known as The Fens was largely reclaimed from this type of vegetation.

*FWAG* – Farming and Wildlife Advisory Group.

*Genus* – a category of closely related species (see below). The first word in a scientific name of an organism refers to its genus, the second to its species, for example all sparrows are *Passer*, but *Passer domesticus* refers only to the house sparrow.

*Gondwana* – the huge southern continent which broke away from the original land mass Panagea. Around 80 million years ago Gondwana broke up into South America, Africa, Madagascar, India, Australia, New Zealand and Antarctica.

*Habitat* – the place in which a plant or animal lives.

*Highest steady density* – the maximal population density of territory-holding males of a dragonfly species at its breeding area. It is expressed as the number of males per 100 metres of water's edge.

*IUCN* – International Union for Conservation of Nature and Natural Resources – the World Conservation Union.

*'Key species'* – a species which is crucial in maintaining an ecosystem (see above).

*National Nature Reserve (NNR)* – a nature reserve of outstanding conservation importance which has been established or designated by the statutory conservation organisations (English Nature, Scottish National Heritage and Countryside Council for Wales) under the provisions of the National Parks and Access to the Countryside Act 1949 and subsequent legislation.

*Nymphaline* – belonging to the butterfly subfamily Nymphalinae, for example small tortoiseshell.

*Open-field system* – the mediaeval system of land tenure, in which large fields were divided into strips. Each farmer farmed a collection of non-contiguous strips. The former position of strips can be observed in ridge-and-furrow fields.

*Palaearctic region* – this large faunal region consists of Europe, Asia north of the Himalayas, North Africa and much of Arabia. Many species (for example the blackbird) are widely distributed in the region.

*RSPB* – Royal Society for the Protection of Birds.

*Species* – a population of interbreeding plants or animals that usually produce fertile young.

*Teneral* – the condition of an adult (or larval) insect soon after moulting when the cuticle is almost colourless and is still unhardened.

*Transect* – a line selected within a study area along which the number and distribution of plants and/or animal species are recorded.

*UNEP* – United Nations Environmental Programme.

*Wash, Washland* – land along the borders of a Fenland river or watercourse which can be flooded to prevent the inundation of adjacent croplands.

*Wildlife* – undomesticated species of plants and animals.

## METRIC CONVERSION TABLES

| Linear | | Area | |
|--------|--------|--------|--------|
| 1 inch | 2.54cm | 1sq inch | 6.452 sq.cm |
| 1 foot | 0.305m | 1sq foot | 0.093 sq.m |
| 1 yard | 0.914m | 1sq. yard | 0.836 sq.m |
| 1 mile | 1.609km | 4840 sq yds | 1 acre |
| | | 2.471 acres | 1 hectare (10,000 sq.m) |

# INDEX

This index covers Chapters 1 to 15, the Postscript and Appendix 1. In this book, Swavesey Nature Reserve refers to my private nature reserve in Boxworth End, Swavesey. Note that the far more important Mare Fen Local Nature Reserve is also in the parish of Swavesey. References to parts of Swavesey Nature Reserve, for example Main Plantation and Old Enclosure, are given as subheadings of Swavesey Nature Reserve. The scientific names of all species of flora and fauna mentioned in the text are given in Appendix 3. Plates listed here are to be found in the pages following page 84.

shelduck 20
shelter belts, see Swavesey
  Nature Reserve, Boundary
  Shelter Belt; Swavesey
  Nature Reserve, Mere
  Shelter Belts
short termism 99–101
shrew, common 59
  pygmy 60
  water 60
silver Y 71
skimmer, black-tailed 72,
  73, 78
skipper, Essex 64, 70,
  Pl. 33
  large 29, 31, 64, 69
  small 29, 31, 64
skylark 19
slugs 80
snail, great pond 61
  water 82, 111
snails 80
snake, grass 60–1, 62, 95,
  107
snake flies 80, 82
snipe 20, 31, 58
  jack 107
snowflake, summer 40, 41
soils 21
  see also clay
sparrow, house 14, 48, 55,
  78
  tree 55
sparrowhawk 55, 56
spearwort, greater 50
species, abundant 96
  aggressive plant 92
  distribution on modern
    farms 89
  interdependency 95–97,
    100–5
  lists 38 (Table 1), 39
    (Table 2), 43 (Table 3),
    44 (Table 4), 51
    (Table 5), 66–7 (Tables
    6,7), 74 (Table 8), 76
    (Table 9)
  number on Swavesey
    Nature Reserve and
    Wicken Fen 81
  rare 87, 96

unidentified 80–81
unrecorded 80–81
valuable 96
see also identification of
  species; and individual
  species
speckled wood 64, 65
  (Fig. 9), 68 (Fig. 10), 69,
  70, 106, Pl. 34
spiders 80
spinney 24
spoonbill 58 (Fig. 8), 59
springtails 80
squirrel, grey 60
St Neots Ware 21
starling 55
Steele, Dick 36–7
stickleback 62
stoat 60
strategy 87–90, 91–2, 93–4
  see also Future Care
straw, barley 111–12
subsidence 30
survival rates, aquatic plants
  50–2 (Table 5)
  trees 38–9 (Tables 1,2)
  woodland flora 40–2
Sussex 72
swallow 55
swallowtail 71
swan, Bewick's 20
  mute 20, 55
Swavesey 13, 15–20 (Figs
  1–3), 54, 85
  history 19, 20, 21, 24,
    25
Swavesey Fen 19–20, 75
Swavesey Nature Reserve 28
  (Fig. 6), 83 (Fig. 12), 91–2,
  93, 95, 98, 106
  amphibians, colonization
    by 61–2
  biodiversity 80–2, 93, 95
  birds, colonization by
    55–9, 81
  Boundary Shelter Belt 28
    (Fig. 6), 29–30, 36, 55,
    57
  Buckingway Plantation 28
    (Fig. 6), 30, 36, 55, 56,
    57, 59, 64, Pl. 31

butterflies, colonization by
  62–72, 81
Corner Plantation 22–3
  (Fig. 4f), 28 (Fig. 6), 45,
  Pl. 21
dragonflies, colonization
  by 31, 32, 72–80
  (Tables 8,9), 81
habitat creation and man-
  agement 26, 35–6, 38,
  40–2, 45, 50, 52–3, 54,
  62, 70, 81, 95, 109
Main Plantation 28
  (Fig.6), 36–43 (Table 1),
  48, 49, 55–6, 57, 59, 60,
  64, Pls 11–18
mammals, colonization by
  59–60
maps 21, 24–5
Mere Enclosure 22–3
  (Fig. 4f), 28 (Fig. 6),
  52–3, 63, 64–9, 70–1,
  81, 109
Mere Meadow 41, 50,
  52–3, 55, Pl. 30
Mere Shelter Belts 28
  (Fig. 6), 48, 49, Pl. 23
Mere, The 21, 22–3
  (Fig. 4f), 28 (Fig. 6),
  47–53, 54, 55, 57–63,
  72–80, 81, 85, 98, 107,
  109, 112, Pls 22,24–9
Old Enclosure 28 (Fig. 6),
  29, 31, 72, 82, Pls 2–3
Old Plantation 28 (Fig. 6),
  30, 36, 38, 42, 55, 56,
  57, 59, 64, Pls 12,20,32
plants, colonization by
  40–2, 50
pond 30–31, 61
reptiles, colonization by
  31, 60–1
trees, colonization by
  43–5
Winter Mere 28 (Fig. 6),
  31
Swavesey Wood, see
  Swavesey Nature Reserve,
  Main Plantation; Swavesey
  Nature Reserve, Old
  Plantation